Ecological Studies, Vol. 81

Analysis and Synthesis

Ecological Studies

Volume 65
Community Ecology of Sea Otters (1987)
Edited by G.R. van Blaricom and J.A. Estes

Volume 66
Forest Hydrology and Ecology at Coweeta (1987)
Edited by W.T. Swank and D.A. Crossley, Jr.

Volume 67
Concepts of Ecosystem Ecology: A Comparative View (1988)
Edited by L.R. Pomeroy and J.J. Alberts

Volume 68
Stable Isotopes in Ecological Research (1989)
Edited by P.W. Rundel, J.R. Ehleringer, and K.A. Nagy

Volume 69
Vertebrates in Complex Tropical Systems (1989)
Edited by M.L. Harmelin-Vivien and F. Bourlière

Volume 70
The Northern Forest Border in Canada and Alaska (1989)
By J.A. Larsen

Volume 71
Tidal Flat Estuaries: Simulation and Analysis of the Ems Estuary (1988)
Edited by J. Baretta and P. Ruardij

Volume 72
Acidic Deposition and Forest Soils (1989)
By D. Binkley, C.T. Driscoll, H.L. Allen, P. Schoeneberger, and D. McAvoy

Volume 73
Toxic Organic Chemicals in Porous Media (1989)
Edited by Z. Gerstl, Y. Chen, U. Mingelgrin, and B. Yaron

Volume 74
Inorganic Contaminants in the Vadose Zone (1989)
Edited by B. Bar-Yosef, N.J. Barrow, and J. Goldshmid

Volume 75
The Grazing Land Ecosystems of the African Sahel (1989)
By H.N. Le Houérou

Volume 76
Vascular Plants as Epiphytes: Evolution and Ecophysiology (1989)
Edited by U. Lüttge

Volume 77
Air Pollution and Forest Decline: A Study of Spruce (*Picea abies*) on Acid Soils (1989)
Edited by E.-D. Schulze, O.L. Lange, and R. Oren

Volume 78
Agroecology: Researching the Ecological Basis for Sustainable Agriculture (1990)
Edited by S.R. Gliessman

Volume 79
Remote Sensing of Biosphere Functioning (1990)
Edited by R.J. Hobbs and H.A. Mooney

Volume 80
Plant Biology of the Basin and Range (1990)
Edited by B. Osmond, G.M. Hidy, and L. Pitelka

Volume 81
Nitrogen in Terrestrial Ecosystem: Questions of Productivity, Vegetational Changes, and Ecosystem Stability (1991)
C.O. Tamm

Volume 82
Quantitative Methods in Landscape Ecology: The Analysis and Interpretation of Landscape Heterogeneity (1990)
Edited by M.G. Turner and R.H. Gardner

Volume 83
The Rivers of Florida (1990)
Edited by R.J. Livingston

Carl Olof Tamm

Nitrogen in Terrestrial Ecosystems

Questions of Productivity, Vegetational Changes, and Ecosystem Stability

With 39 Figures

Springer-Verlag Berlin Heidelberg New York
London Paris Tokyo Hong Kong Barcelona

Prof. Dr. CARL OLOF TAMM

Stavgårdsgatan 11
161 37 Bromma, Sweden

Department of Ecology and Environmental Research
Swedish University of Agricultural Sciences
P.O. Box 7072
750 07 Uppsala, Sweden

ISBN 3-540-51807-X Springer-Verlag Berlin Heidelberg New York
ISBN 0-387-51807-X Springer-Verlag New York Berlin Heidelberg

Library of Congress Cataloging-in-Publication Data. Tamm, Carl Olof. Nitrogen in terrestrial ecosystems: questions of productivity, vegetational changes, and ecosystem stability / Carl Olof Tam. p. cm. – (Ecological studies; v. 81) Includes bibliographical references (p. Includes index.
ISBN 3-540-51807-X (alk. paper). – ISBN 0-387-51807-X (U.S.: alk. paper)
1. Nitrogen cycles. 2. Biotic communities. 3. Plants, Effect of nitrogen on. 4. Ecology. 5. Forest ecology. I. Title. II. Title: Terrestrial ecosystems. III. Series.
QH344.T36 1990 574.5'2642 – dc20 90-10292 CIP

Typesetting: K+V Fotosatz GmbH, 6124 Beerfelden
2131/3145 (3011)-543210 – Printed on acid-free paper

Preface

The actual work with this book started in 1985, when Dr. J. Nilsson of the Swedish Environmental Protection Board asked me to write down and expand some discussion contributions from Swedish and Nordic meetings on the theme of "critical loads" of sulphur and nitrogen. The discussion concerned primarily atmospheric deposition, but also the effects of nitrogen fertilization were of interest. Somewhat later Dr. Nilsson asked me to prepare a more comprehensive review to be used as one of the background documents of the Nordic/UN-ECE workshop on Critical Loads for Sulphur and Nitrogen, held at Skokloster, Sweden, March 21–24, 1988.

However, there is also a longer history behind the book. I owe much of my interest in the functions of nutrients in terrestrial ecosystems – with nitrogen as a focal point – to my old friends Carl Malmström and Lars-Gunnar Romell, who together guided my first research at the Swedish Forest Research Institute, when I became employed there in 1949. Both these important ecologists had a genuine feeling for what is now termed ecosystem ecology, documented in their writings even before Tansley introduced the word ecosystem in 1935. They were also interested in how old land-use methods had formed the present mosaic of farmland, pastures, and forests, and they both took an active interest in the preservation of examples of the historical, man-made landscape.

If forest nutrition is defined as the study of the relationships between the functions of forest ecosystems and the availability of plant nutrients on their sites, this topic was hardly recognized as a science around 1950. It could not properly be classified within traditional disciplines, being neither classical physiology nor ecology, at the same time as it was too theoretical for silviculture around 1950. In wetland research the distinction had been made between ombrotrophic and minerotrophic types, but there was little understanding of the roles played by different plant nutrients.

However, one aspect of forest nutrient relationships had long been recognized, viz. nutrient cycling in forest ecosystems, with the first studies made already in the late 19th century, as attempts to explain the deleterious effects of litter-raking on Middle European forests. The interest for litter increased again from the 1920s onwards, as it became clear that litter quantity and quality was decisive for humus layer properties, which in turn influenced the rest of the soil profile. Very soon, however, ecosystem nutrient cycling would become a big issue, thanks to P. Duvigneaud, E.P. and H.T. Odum, J.D. Ovington, P.J. Rennie, and the International Biological Programme (1964–1974).

Without much connection with the rising interest in nutrient cycling, field experiments were laid out, intending to stimulate tree growth, often in connection with afforestation or replanting. It was not believed that the forest products should ever pay back the costs of the expensive commercial fertilizers used in agriculture, so most forest experimenters preferred cheap materials such as lime, ground basaltic rock, or slag products from steelworks. The very first experiments date from around the turn of the century. The initiative often came from practical foresters, even if also scientists started to make systematic experiments, e.g. Nemec in Czechoslovakia, Süchting in Germany, Hesselman, Romell and Malmström in Sweden, Mitchell and Chandler and later Heiberg and co-workers in the north-eastern United States.

The results of these early experiments, as reported by White and Leaf (1957), were too dispersed to allow general conclusions. We may also assume that the published reports constitute a biased "sample", as many negative results were never published. In many cases the interpretation was complicated by the lack of statistical design or the use of poorly defined material (wood ashes, slags, or ground rock). Yet there were striking examples of successful application of phosphates or other mineral material in heathland afforestation, on poor sands and on drained peatlands, and of nitrogen in Sweden and NE United States. Some of the experiments where lime had been applied were also successful, particularly when the lime had been mixed into the soil.

At the same time foresters and biologists had collected vast amounts of information on the importance of various site factors for plant growth and distribution. It was a growing understanding that many of the edaphic factors regulating plant distribution were connected with the supply of plant nutrients (acidity as related to lime content; humus forms; wetland differentiation from ombrotrophic bogs to intermittently flooded valley bottoms, etc.).

The apparent lack of couplings between these areas – nutrient cycling studies, experimental work with nutrients, and plant distribution studies – was a challenge for a young scientist. Much of the information collected in this book, both from the literature and from investigations by myself and many collaborators, has a link to my lifelong search for bridges between these areas.

Since the early 1950s much has happened in ecosystem ecology, with an almost glamourous period culminating with the International Biological Programme (1964–1974). Even if evolutionary biology claimed to represent the true scientific frontier, the ecologists could not ignore the alarm signals during the 1960s from Rachel Carson, Svante Odén, and others who revealed connections between pollution problems and ecosystem processes such as food chain enrichment and nutrient losses from sites. At the same time it became clear that several types of pollution problems could no longer be considered as local problems, but were in fact global or at least concerned large regions. It became very clear that society cannot cope with environmental problems without a good understanding of ecosystem functioning. This again requires cooperation between many different disciplines, something for which terrestrial ecologists were not well trained. In contrast, limnologists had used a more interdisciplinary approach already from the start of their science in the beginning of the century.

However, all questions cannot be answered by new, large and integrated projects. The time factor has always been important in ecological research, not least in such dealing with forests. There are many timelags in processes and regulatory mechanisms in a forest ecosystem. Therefore it is extremely important to extract as much information as possible from work already done. Although an increasing number of scientists now understand the need for looking at old data in a new and more integrated way, it is my feeling that this approach still offers new and interesting possibilities.

The last two decades of research in "chemical climate changes" have made it clear that the global cycles of carbon, sulphur, and nitrogen have been affected by anthropogenic disturbances in a much more profound way than earlier believed. There have been attempts to collect evidence of the ecological effects of the changes, notably by SCOPE (Special Committee for Problems of the Environment of the International Council of Scientific Unions, ICSU) in cooperation with several governmental and nongovernmental international organizations. Despite these efforts – without which this book could not have been written – there seems to me to be a need for syntheses, connecting the development of various landscape elements, all with their own history of natural succession and human impacts, with the present pollution situation.

This book is focused on the ecological and biogeochemical behaviour of nitrogen in natural and semi-natural terrestrial ecosystems, where this element often plays a key role in the regulation of ecosystem processes. Industrialization and modern agriculture are now adding important contributions to the global nitrogen cycle. Regionally the anthropogenic emissions dominate over the natural ones. Strong changes have already taken place in many ecosystems, and further changes can be foreseen. Even if emission of nitrogenous compounds is only one of many pollution problems threatening terrestrial ecosystems, it seems worthwhile to try to synthesize existing information in the field and to draw some conclusions on the future development of these ecosystems, when both land-use practices and emissions of atmospheric pollutants change.

The literature on nitrogen in terrestrial ecosystems is very large, so the emphasis has been given to the information most relevant for the integrated look attempted here, and there may certainly be omissions even in this respect. A considerable part of the information in Chapter 2 can also be found in textbooks on plant physiology and soil microbiology, and thus is not new for readers well familiar with these disciplines. Yet it is the author's belief that many readers might valuate a short repetition of those traits of nitrogen metabolism in plant and soil that are most important in connection with ecosystem functions and with ongoing environmental changes. There are also a number of references hopefully useful to those readers who are particularly interested in such aspects of nitrogen metabolism that are only briefly discussed in this book. An example: as the central theme here is effects of nitrogen compounds on ecosystem processes, questions on how ecosystem processes affect the production of nitrogenous trace gases are not dealt with in much detail, despite the importance of these gases in atmospheric chemistry and global climate research.

As the nitrogen cycle in intensively cropped agroecosystems is regulated largely by the supply of nitrogenous fertilizers, and by the harvesting of nitrogen with the crops, their ecology is a bit outside the central theme of this book. Therefore literature on agroecosystems has only been quoted for comparisons with conditions on less intensively used land. Agroecosystems are often better studied than natural ecosystems, and therefore such comparisons can be very useful, particularly when thinking of possible scenarios for the future.

Some readers may find the book biased in the sense that many cases discussed are taken from the northern coniferous zone. The reasons for this are not only the author's acquaintance with the boreal forest, but also that this biome provides many striking examples of ecosystems limited by nitrogen supply. Much of the early work on forest and tree nutrition was done in the northern coniferous forest and adjacent parts of the north temperate zone, both in Europe and in North America, and consequently much of the pioneering field experimentation with nutrients on forest sites was made in northern Europe and in North America. The fundamental physical, chemical and biological laws which govern ecosystem processes are the same everywhere, but the proportions of energy and matter taking different pathways may vary greatly according to climatic and edaphic factors. Consequently, the relative roles of various factors limiting growth and influencing other ecosystem functions will be different in different biomes. Yet the similarities are often more important than the differences. It is the belief of the author that the reacting to changes in nitrogen availability of northern forests will be of interest also to students of other nitrogen-limited ecosystems. In e.g. tropical forests other factors may be in shorter supply than nitrogen in the natural state, but nitrogen limitation is often induced by human operations, such as shifting cultivation or plantation forestry.

Nobody would be happier than the author, if this book became quickly outdated, either because some of the views taken can be shown to be too pessimistic, or because there will be a drastic reduction of emissions of nitrogen compounds and other air pollutants in the near future.

As is clear from the above, this book could not have been written without help in many ways from friends and colleagues. Among those who first helped me to understand ecological principles I have already mentioned Carl Malmström and Lars-Gunnar Romell, but my first teacher in ecology and physiological botany, Gottfried Stålfelt, was another scientist who warned of the misuse of natural resources already in the 1940s and 1950s. My father, Olof Tamm, was not only a pioneer in soil science, but also passed on to his students and to his son much of his deep knowledge of forest-soil relationships.

Different versions of the manuscript or parts of it were read by several colleagues, who all gave useful comments. Further colleagues provided answers to difficult questions and helped to evaluate references for topics less familiar to me. However, the responsibility for errors or oversimplifications which may still occur in the text, rests entirely with the author. For their help I wish to thank Drs. Aron Aronsson, Björn Berg, Hermann Ellenberg, Stig Larsson, Ingvar Nilsson, Jan Nilsson, Jerry S. Olson, Thomas Rosswall, and Germund

Tyler. Dr. Olson also suggested that a final version of the draft he read should be published in the Series Ecological Studies by Springer-Verlag. I am extremely grateful to the staff of Ecological Studies for their careful editorial work. The economic support provided by the Swedish Environmental Protection Board for the early version of the manuscript is gratefully acknowledged.

Many colleagues have kindly given their consent to my use of their Figures and Tables, as evident from the source notation in Figure or Table legends. Drs. A. Aronsson, S. Linder and B. Popovic have constructed some of the diagrams taken from manuscripts as yet unpublished. Other material coming from planned publications by other Swedish colleagues is reproduced with their consent. My sincere thanks are due for all this generosity. In several cases the permission to reproduce material was granted by publishing companies, journals, or organizations, as listed at the end of the book.

The literature search ended in March 1989, but occasional references were introduced later.

Uppsala, Spring 1990 Carl Olof Tamm

Contents

1 Introduction: Geochemical Occurrence of Nitrogen. Natural Nitrogen Cycling and Anthropogenic Nitrogen Emissions

Together with carbon, oxygen and hydrogen, nitrogen is one of the four most common elements in living cells and an essential constituent of proteins and nucleic acids, the two groups of substances which can be said to support life. Yet the element is not particularly common on earth, with the exception of the atmosphere, which contains almost 80% nitrogen. The estimated 11000 to 14000 teragrams (10^{12}) nitrogen in living biomass (mainly terrestrial plants) is equivalent to about three parts per million of the atmospheric nitrogen (Table 1.1). Other important nitrogen pools are: soil organic matter, rocks (in fact the largest single pool), sediments, coal deposits, organic matter in ocean water, and nitrate in ocean water. The next most common gaseous form of nitrogen in the atmosphere after molecular nitrogen is dinitrogen oxide.

The size of pools does not indicate anything about the dynamics of the annual global fluxes of nitrogen between the more important pools. Several of the estimates, in particular of the fluxes, are calculated more or less indirectly and are rather uncertain. However, in most cases the order of magnitude and the proportions between various pools and fluxes are more important than exact figures. Figure 1.1 indicates that anthropogenic contributions of nitrogen compounds (ammonia and nitrogen oxides mediated by human technologies) to the global cycle cannot be neglected. The biological nitrogen fixation is still by far the largest input (140 teragram N year^{-1}) to terrestrial ecosystems, but anthropogenic inputs and outputs are increasing. Söderlund and Svensson (1976) estimated ammonia flow to the atmosphere from coal burning at 4–12 teragram N year^{-1} and from domestic animals and human beings at 20–35 teragram, far more than from wild animals. Anthropogenic emission of NO_x-nitrogen by combustion was calculated at 19 teragram for the year 1970, but this figure is continuously rising (Fig. 1.2). So is the use of nitrogen fertilizers, "anthropogenic nitrogen fixation" (Fig. 1.3), even if it is still small compared with the biological nitrogen fixation on a global scale.

However, comparisons between the different global fluxes in Fig. 1.1 do not tell the whole story. Even if the anthropogenic fluxes are smaller than the global ones, they are concentrated to industrial and intensively farmed regions, and there they often exceed the natural fluxes, sometimes by orders of magnitude. The residence time of ammonia and most nitrogen oxides (not dinitrogen oxide) in the atmosphere is relatively short under common weather conditions, so much of the deposition will occur within the region of emission or at least within distances of one to a few thousand kilometres.

Table 1.1. Global inventories of nitrogen in the terrestrial, oceanic, and atmospherics systems (Tg N). Quoted from Holdgate and White (1977), data compiled by Söderlund and Svensson (1976)

System	Amount
Terrestrial	
Plant biomass	$1.1-1.4\times10^4$
Animal biomass	2×10^2
Litter	$1.9-3.3\times10^3$
Soil	
Organic matter	3.0×10^5
Insoluble inorganic	1.6×10^4
Microorganisms	5×10^2 [a]
Rocks	1.9×10^{11}
Sediments	4×10^8
Coal deposits	1.2×10^5
Oceanic	
Plant biomass	3.0×10^2
Animal biomass	1.7×10^2
Dead organic matter	
Dissolved	5.3×10^5
Particulate	$0.3-2.4\times10^4$
N_2 (dissolved)	2.2×10^7
N_2O	2.0×10^2
NH_3	0.9
NO_3^-	5.7×10^5
NO_2^-	5.0×10^2
NH_4^+	7.0×10^3
Atmospheric	
N_2	3.9×10^9
N_2O	1.3×10^3
NH_4^+	1.8
NO_x	$1-4$
NO_3^-	0.5
Org. N	1

[a] This amount is inevitably included in the figure for organic nitrogen in the soil.

Since the 1950s, measurements of wet deposition of nitrogen compounds have been available in Europe, although with somewhat varying accuracy. Single longer series occur, e.g., at Rothamsted in England, where inputs between 4.1 and 5.0 kg N ha^{-1} $year^{-1}$ were reported from 1888 to 1913 and between 15.1 and 18.2 kg from 1969 to 1978 (Anonymous 1983). The quality of the old data, particularly for ammonia nitrogen, is often questionable, because of both inaccurate methods and risks for contamination, which makes it difficult to form a general picture (Brimblecombe and Stedman 1982; Skeffington and Wilson 1988). Yet there seems to be little doubt that the wet deposition of ni-

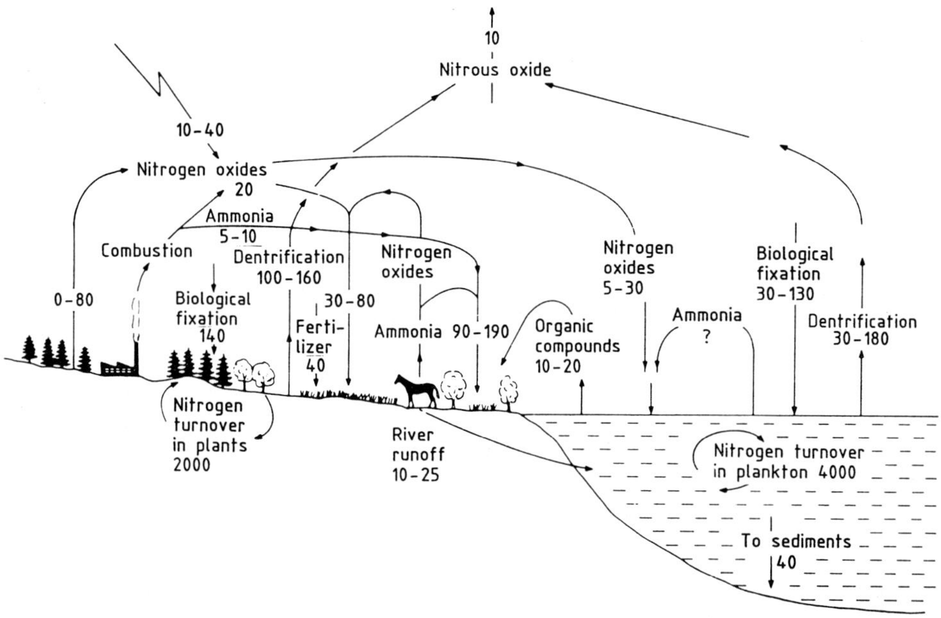

Fig. 1.1. Global fluxes of nitrogen in teragram (10^{12} g) per year (Rosswall 1983)

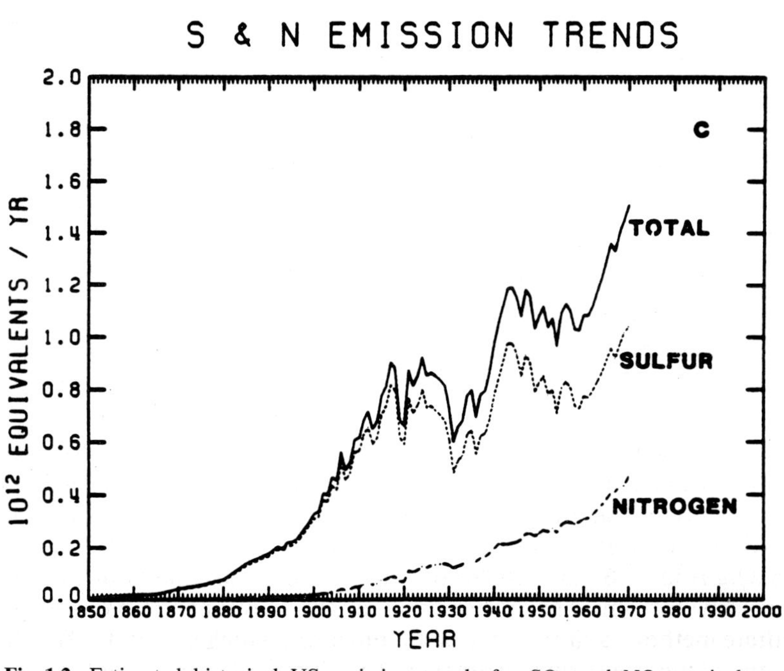

Fig. 1.2. Estimated historical US emission trends for SO_x and NO_x equivalents (Husar and Holloway 1983)

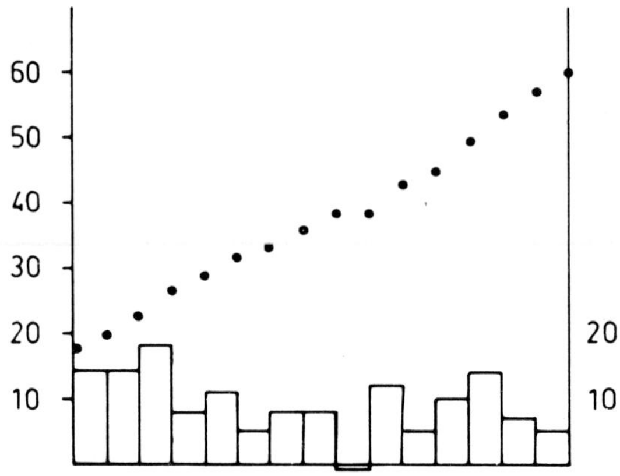

Fig. 1.3. The global yearly utilization of nitrogen fertilizers (1965/66–1980/81). *Dots* mark consumption in teragrams N (*left scale*) and *bars* yearly percentage change (*right scale*). Compiled by Rosswall 1983, data from FAO

trate has increased from about 1860 to 1960 (Skeffington and Wilson 1988), and that in large parts of Europe there is a more recent increase in both ammonia and nitrate nitrogen, roughly parallel to the increase in industrial combustion, motor traffic, and agricultural nitrogen cycling. Although dry deposition is more difficult to measure, there is reason to believe that it follows a similar pattern.

So far most attempts to measure increases in nitrogen deposition in some kind of biological or geochemical archive have given inconclusive evidence. The exception is inland ice cores, where studies (Neftel et al. 1985) indicate a doubling of nitrate concentrations from 1896 to 1978 in south Greenland. Other possibilities to be considered are ombrotrophic bogs, tree rings, and comparisons between old and new plant analyses.

Geochemical profiles in ombrotrophic bogs often show higher concentrations of nitrogen in the upper few decimetres than lower down (Mattson and Koutler-Andersson 1954, and others) as well as of various mineral elements. This enrichment could result from increased atmospheric supply, but it could also be a consequence of ecosystem retention of essential nutrients (roots pumping nutrients back into the plants and returning them in litter on top of the bog). For some elements, including nitrogen, an increase in decomposition leads to increases in concentrations (Heinselman 1979; Malmer and Holm 1984). Still, careful studies of bog chemical stratigraphy related to bog growth might give valuable information.

Annual rings in tree trunks can give information as to the chemical environment when the ring was formed (Lepp 1975), but the method works best with elements of low mobility in the plant. Element mobility in living organisms and dead organic matter varies within a wide range from very immobile lead (see e.g., Tyler 1978) to mobile potassium and phosphorus. Nitrogen is even more mobile than the two elements mentioned last and is partly withdrawn from the sapwood when it turns to heartwood. Nitrogen is also often a growth-limiting element, which means that variations in nitrogen availability in many cases may cause variations in ring width rather than in wood nitrogen concen-

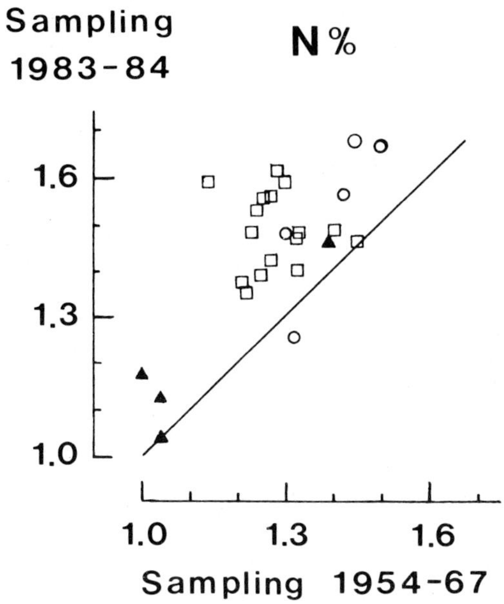

Fig. 1.4. Concentrations (percent dry weight) of nitrogen in exposed current spruce needles, sampled in autumn from the same forest plots on different occasions. *Squares* Tönnersjöheden Experimental Forest, SW Sweden. *Circles* Other sites in south Sweden (Götaland). *Filled triangles* sites in middle and north Sweden (Aronsson 1985a)

trations. One attempt to relate pine heartwood nitrogen concentrations to changes in nitrogen availability on the site was not very successful (Popovic 1966). For similar reasons as for tree rings, the prospects do not look very promising for using other biological archives with annual rings (e.g., land-snail shells, Gärdenfors et al. 1988) for tracing changes in nitrogen availability in the past, while such studies might be extremely valuable in the case of less mobile elements.

Comparative analyses of plant material collected in different periods require strict control of both the sampling and the site to give informative results. Such control is possible around local emissions. Studies in California along air pollution gradients have shown gradients in foliage nitrogen concentrations and nitrogen/phosphorus ratios (Zinke 1980). However, as there is considerable variation in chemical composition of plant organs such as needles and leaves, i.a., between positions in crown, seasons, and years, historical data seldom satisfy the requirements for reproducibility. Yet it is of interest that a pilot study by Aronsson (1985a) of needles sampled in various parts of Sweden around 1960 and in 1983 and 1984, from the same stands and in many cases from the same trees (in permanent forest plots), shows systematic differences in element concentrations between needle samplings, i.a., higher needle nitrogen concentrations in southern Sweden but not in northern Sweden (Fig. 1.4). There is no evidence at hand to show that the ageing of the trees per se would lead to higher nitrogen concentrations. However, since both old and new samplings represent short time series, in some cases single years, between-year variation may be of importance. Other historical comparisons have shown differences in needle concentrations over time of other elements, e.g., magnesium (Hüttl and Zöttl 1985).

If it thus must be concluded that there is meagre direct evidence of changes in nitrogen concentrations and amounts in forests and other natural ecosystems, related to increased atmospheric deposition, there is much indirect evidence. As will be discussed in a later section, both terrestrial and aquatic nitrophilous plant species have increased in large regions, with consequences both for other ecosystem components and for functional relationships.

2 Nitrogen in Plants and in Soils: Physiological and Microbiological Background for Biological Nitrogen Turnover

2.1 Nitrogen in Plants

Most plants and other living organisms need nitrogen in larger amounts than they need essential elements other than carbon, oxygen, and hydrogen. Nitrogen is a major and essential constituent of living cells. The proteins are polymerized amino acids, and nucleic acids are also polymers containing nitrogen in their constituents. In plants the composition of cytoplasm is relatively constant, while other constituents such as cell wall substances, energy-rich substances (starch), and so-called secondary metabolites may vary widely with plant organ, age, growth conditions, and species. There is often a close relation between the amount of nitrogen available to roots and total plant biomass in the ecosystem, which can be traced back to the fundamental relation between available nitrogen and plant cytoplasm. Animals generally show less variation in tissue composition than plants, but are equally dependent on a steady supply of nitrogen, preferably in the form of proteins or amino acids.

As nitrogen is a constituent of chlorophyll and of enzymes participating in the photosynthesis, the chlorosis often observed in severely nitrogen-deficient plants is often taken as evidence that a direct relationship must exist between leaf nitrogen concentration and photosynthetic efficiency. Such a relationship has also been demonstrated in nitrogen-limited systems on the single leaf level (Field and Mooney 1986) over a wide range of life forms and environments (Fig. 2.1).

On the stand level, conditions may be more complicated. Linder and Troeng (1980; Fig. 2.2), found a 15% to 20% increase in rate of light-saturated photosynthesis per unit area of leaf, comparing irrigated-fertilized *Pinus sylvestris* with trees irrigated only. Yet the annual canopy photosynthesis was doubled, so it was concluded that allocation of resources to increased needle growth played an important role for the response of the whole trees (Fig. 2.3; Linder 1987, with further references). The amounts of nitrogen available at any given moment in a terrestrial ecosystem are often limited. It is characteristic of nitrogen that only a small fraction of the total amount in a terrestrial ecosystem occurs in inorganic form (mainly as ammonium or nitrate ions), the form in which nitrogen is normally available to higher plants. Mycorrhizal fungi may also take up certain quantities of low-molecular organic nitrogen in, e.g., amino acids (Alexander 1983; Read 1983; Nylund 1988). There is a continuous decomposition of nitrogen-containing organic matter in the soil. Mineral nitro-

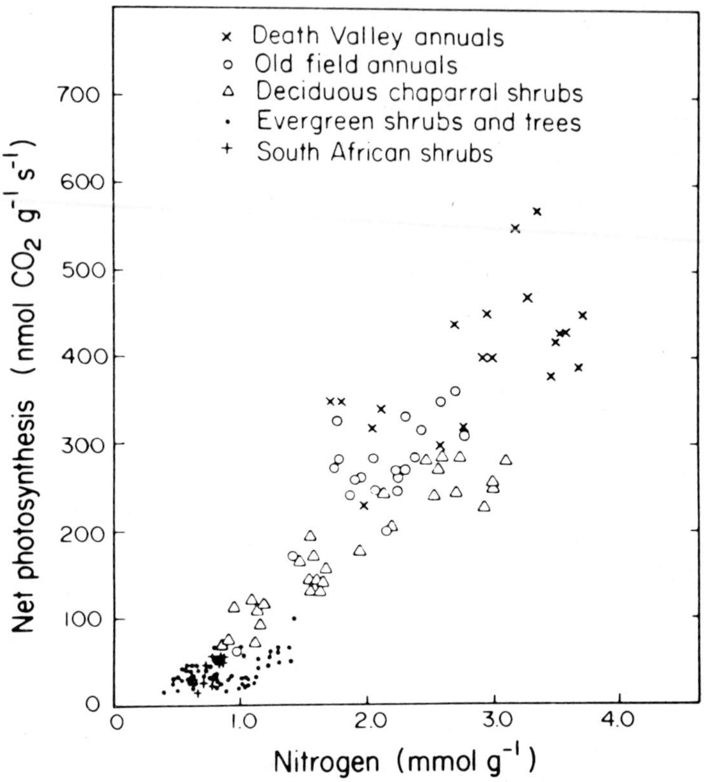

Fig. 2.1. Photosynthetic capacity, A_{max}, plotted against leaf nitrogen content, both expressed on a leaf weight basis, over a range of life forms and environments. A_{max} is the rate of net photosynthesis, measured at light saturation, optimum temperature, relatively high humidity, and normal carbon dioxide concentration (Field and Mooney 1986)

gen is released and then rapidly taken up again by roots and microorganisms and again transferred to organic form. Soil concentrations of ammonium or nitrate ions are therefore not good expressions for the availability of nitrogen to roots, when different ecosystems are compared (Ellenberg 1977; Keeney 1980; Vitousek and Andariese 1986).

A theoretically better measure is the nitrogen flux density, the amounts of nitrogen made available per unit of soil per unit of time (Ingestad 1982). However, this must be estimated indirectly, except in simple artificial systems where nitrogen uptake can be measured over a period of time (see references in Chapin et al. 1986). A consequence of the nitrogen flux density concept is that a plant with a high growth rate needs a higher flux density (in solution culture a higher relative addition rate, Ingestad 1987) than a plant with a lower growth rate. This applies both to different species, e.g., r-selected and K-selected ones (generally higher growth rate in r-selected species, if other circumstances are similar) and to different age stages in the same species. Tree seedlings and sap-

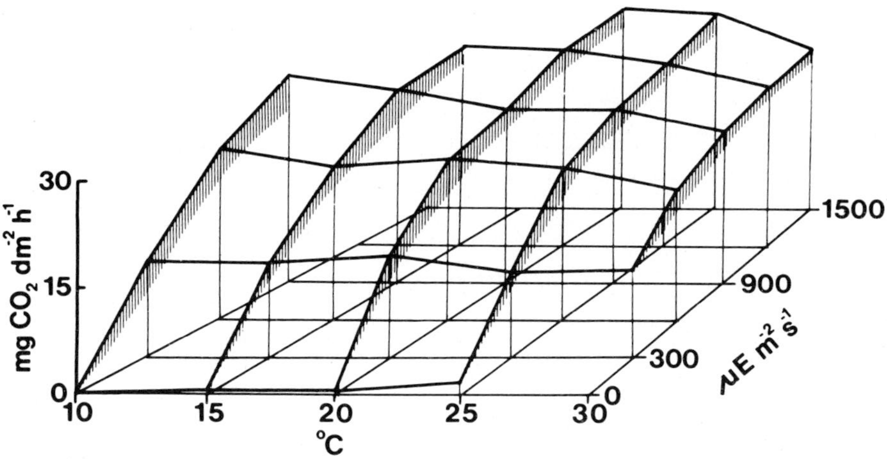

Fig. 2.2. Light and temperature response of net photosynthesis for a current shoot of Scots pine on irrigated and irrigated/fertilized plots. The *shaded* area shows the difference between the treatments in photosynthetic rate per unit area of needles where the irrigated/fertilized plots always had the higher rates. The diagram is based on measurements from August 1975 (Linder and Troeng 1980)

lings cannot survive and develop in to adult trees unless they have a higher relative growth rate than the older trees. Consequently they need a high nutrient flux density. In most cases tree seedlings cannot compete successfully with the adult trees of the same species, when growing under a canopy. This is usually explained by competition for light and sometimes water ("big trees have deeper roots"), but trenching experiments, such as those done by Toumey (1928) and Hesselman in the 1920s (Romell and Malmström 1945), convincingly showed that tree seedlings and other plants can grow also beneath a canopy if they are relieved of root competition, i.e., at a nutrient flux density to their roots other than in the presence of the roots from the old trees. Field layer vegetation in a nitrogen-limited ecosystem reacts upon trenching in a manner very similar to the reaction upon nitrogen fertilization (Figs. 2.4, 2.5). With some delay tree saplings also increase growth (see also Sects. 3.3, 4.4).

A characteristic property of nitrogen in the plant, or of the plant's way of handling nitrogen, is the high mobility of this element, which can easily be redistributed between different organs. When a tissue becomes senescent, much nitrogen is often withdrawn from it to growing tissues or storage organs, including seeds. The degree of recovery of nitrogen from senescent tissues is influenced by both external and internal factors. If nitrogen occurs in excess in the environment, a higher proportion of the nitrogen content may stay in the litter. But the life form of the plant species also plays an important role. Annual plants can only store nitrogen in seeds, while other life forms have further possibilities, such as storage in roots, rhizomes, and above-ground organs. It is clear that effective mechanisms for redistribution of scarce nutrients as well

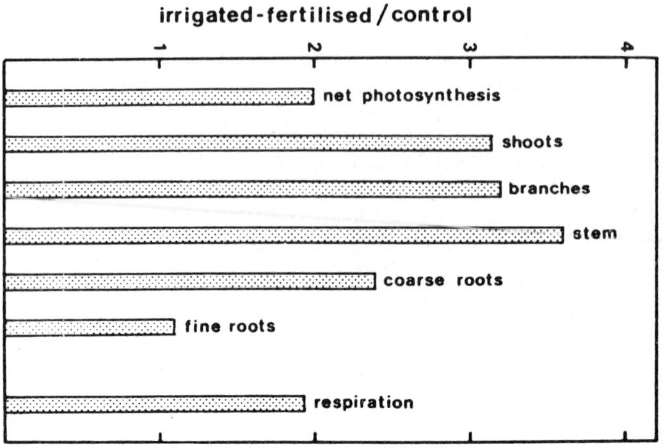

Fig. 2.3. Changes in growth allocation in response to irrigation and fertilization in a young stand of *Pinus sylvestris* on a poor site, with growth naturally limited by nitrogen supply. The *bars* represent the ratios between the means for the irrigated and fertilized and for the control plots, in terms of annual photosynthesis and annual production of various biomass components (Linder and Axelsson 1982)

as energy-rich substances give a plant a competitional advantage in a poor environment. Yet different life forms often coexist in nutrient-stressed environments, so there is not necessarily one single solution to the problem of how to handle nutrients most efficiently (Shaver and Chapin 1980).

One of the more successful solutions to the problem of conserving nutrients for future growth and survival is that adapted by evergreen trees and dwarf shrubs of the northern coniferous forests (Waring and Franklin 1979; Chapin and Kedrowski 1983). A larger part of the stores of easily available carbohydrates in these species is found in the foliage (Tamm 1955; Rutter 1957; Bryant et al. 1983), which may live for several years or even decades, with annual fluctuations in carbohydrate content (Ericsson 1978, 1979). Also, nitrogen and other mobile nutrients such as phosphorus are stored in the needles and translocated to new growth (Tamm 1955). The degree to which stored starch is translocated out of the needles for the production of new organs appears to depend on the nutrient level in the tree. Better nutrient conditions apparently create more efficient sinks for starch, in agreement with changes in resource allocation described earlier (Ericsson 1979).

The main disadvantage of overwinter storage in needles seems to be the risk for needle losses due to climatic stresses during winter, which sets northern and altitudinal limits for the survival of coniferous trees. Evergreen dwarf shrubs covered by snow in winter also occur above the tree line and on the tundra.

Moist tropical forests represent another type of site with low nutrient availability in the soil. It is therefore interesting that trees in this environment may translocate back about half their leaf content of nitrogen and phosphorus before litter formation (Vitousek and Sanford 1986).

Fig. 2.4. The effect of removal of competition from tree roots and dwarf shrubs on the vegetation in an old spruce forest at Hasberget, Orsa Finnmark, central Sweden. All tree roots were cut in July 1936 and again in September 1939 around several microplots. Steel plate isolation was installed in September 1939 around the microplots, as shown in the picture. The dwarf shrubs (*Vaccinium myrtillus*) were cut in August 1936 and again in June 1938. The grass *Deschampsia flexuosa* soon dominated within the trenched plots, but *Rubus idaeus* and *Luzula pilosa* occurred as well. Photo by LG. Romell, September 1940

It is clear from work on the genetics of agricultural plants that part of the variation in the allocation of resources in plants is genetically controlled. The importance of this aspect in connection with nitrogen nutrition has not yet been given much attention with regard to forest trees and wild plants. A few other mineral nutrients (particularly phosphorus and potassium) are also redistributed efficiently when they are in low supply, but most other bioelements remain to a large extent in the organ where they were first deposited after uptake.

The high mobility of nitrogen within the plant enables it to use the nitrogen taken up in a very efficient way, using it over and over again. Together with

Fig. 2.5. The effect of additional nitrogen supply (irrigation with ammonium nitrate solution every other week during summer, to the *left* in picture) on field layer vegetation in old mixed pine-spruce forest at Storliden, Kulbäcksliden Experimental Forest, N Sweden. The total amount of nitrogen was 540 kg ha^{-1} over the 3 years of treatment, most of which had been applied before early August 1946, when the photo was taken. Note shift in dominance from *Vaccinium myrtillus* to *Deschampsia flexuosa*. The reaction was similar on all plots receiving ammonium nitrate, irrespective of whether phosphoric acid or wood ashes were applied as well. (For further details, see Malmström 1949; photo LG. Romell)

the relative scarcity of available nitrogen in the soil, the high nitrogen use efficiency (Ingestad 1977; cf. Vitousek 1982) is part of the explanation of the fact that there is so often a good relationship between the nitrogen availability in the ecosystem and its net primary production, as will be discussed later. Chapters 3 and 4 will be devoted to a description of how nitrogen-depleted and nitrogen-enriched ecosystems function, so here we shall only describe briefly some of the physiological processes which change when nitrogen supply increases above a point where plant growth is no longer stimulated.

It should be mentioned that high nitrogen use efficiency is by no means a characteristic restricted to vascular plants growing on sites with low levels of available nitrogen. Wood-destroying fungi live in an environment very low in available nitrogen and have physiological adaptations permitting good development under these conditions (Levi and Cowling 1969). *Sphagnum* moss is another example of plants adapted to environments low in mineral nitrogen, such as ombrotrophic bogs, and they are able to survive on the normally very low supply of nitrogen from the atmosphere, only under some conditions supplemented with nitrogen fixed by associated blue green algae. Labelled carbon and phosphorus are redistributed from older parts to the apex of *Sphagnum recurrum* (Rydin and Clymo 1989) and there is no reason why nitrogen should not follow the same pattern. When the amounts of nitrogen compounds from emissions increase above a certain level, as in the Pennines in northern England, the *Sphagnum* loses its ability to reduce and utilize nitrate, at the same time as the entire vegetation structure is damaged (Lee and Woodin 1988).

Even if the physiological need for nitrogen is satisfied, plant roots continue to absorb ammonium and nitrate ions. Ammonium ions are rapidly metabolized, a process requiring comparatively little energy, as the redox state of nitrogen remains unchanged (Pate 1983). As the ammonium nitrogen is transferred to amino acid or amide nitrogen, cell metabolism must provide the organic acids necessary for this process. The acids in question are produced from carbohydrates in normal metabolic processes, as long as the cell has enough carbohydrates in storage. The most common intermediary products formed from ammonium ions and organic acids are glutamine and asparagine, which serve both for translocation and for temporary storage of nitrogen in many plants.

Besides being a pathway for nitrogen utilization, the amination means a detoxication, as free ammonia and also high concentrations of ammonium ions have toxic effects in living cells. In the root medium a secondary effect of high ammonium concentrations may be acidification. Ammonium ions are more rapidly absorbed by roots than most other cations and anions, because the amination creates an effective sink within the root. As most cation uptake occurs in exchange for hydrogen ions, a local acidification takes place, which under some circumstances may damage the roots (Olsen 1921). Whether this acidification is of importance in a wider context depends on the source of the ammonium ions. If these are formed within the ecosystem by normal decomposition and mineralization processes, the uptake will not affect the proton budget of the ecosystem. On the other hand, if they are added from outside with fertilizers or pollutants, the net result depends on what happens with the accompanying anion, if any. If the anion (usually nitrate or sulphate) leaves the soil profile with the percolating water, there is a net acidification. In the case of urea added with fertilizer or farmyard manure, the immediate reaction in the substrate is hydrolysis of urea, implying alkalinization, which may be compensated for if the ammonium ions are taken up by organisms. If a larger part of the ammonium ions from the urea are nitrified, the result is acidification.

Nitrate ions are also easily taken up by roots, although at a higher energy cost than ammonium ions (Pate 1983, 1986). They are exchanged for bicarbonate or hydroxyl ions, which leads to a counteraction of the acidification caused by cation uptake. Upon entrance into the root two things can happen: (1) rapid reduction of the nitrogen and formation of amino acids or amides, as above, or (2) translocation of the nitrate to other parts of the plant, including the leaves. The nitrate is not very toxic, and a reduction to amino nitrogen may take place in green organs, with a coupling to the photosynthesis, a pathway which seems to require less energy than reduction in the dark (Gutschick 1981; Pate 1983; Smirnoff and Stewart 1985). The temporary accumulation of nitrate in the leaves or other organs (taproots, petioles, which may hold higher nitrate concentrations than leaves) and then reduction coupled to photosynthesis is a characteristic of certain plants, while others normally reduce all nitrate immediately upon entrance into the plant. Determination of the enzyme nitrate reductase in plant leaves has become a useful indirect method to assess soil nitrification (Havill et al. 1974; Lee and Stewart 1978; Högberg et al. 1986; Gebauer et al. 1988), but cannot, according to what has been said, be used for all species.

"Nitrate plants" (Hesselman 1917) are many common crop plants such as most cereals, tomato, spinach, lettuce, as well as many weeds. Under natural conditions some nitrate plants, e.g., *Chenopodium* species and *Beta maritima*, have their habitats on sea-shores. Other nitrate plants such as *Epilobium angustifolium* (fireweed), *Rubus idaeus*, and *Rumex* and *Senecio* species occur on disturbed sites such as after a forest fire. Not all of these species require nitrate to develop well (Tamm 1956), but well-growing specimens on the types of sites described almost invariably contain nitrate in their leaves. In other plant species, such as our common forest trees and practically all plants characteristic of later successional stages in the boreal forest (e.g., *Vaccinium* species), nitrate is not found in the leaves under normal conditions, not even on nitrate-rich sites. Some species may even lack the ability to take up nitrate (Lee and Stewart 1978).

As seen from the discussion above, plants can store excess nitrogen in two ways, either as organic compounds (glutamine, asparagine, nitrogen-rich amino acids such as arginine), or as inorganic nitrate nitrogen. This is not the place to speculate on the various ecological implications of these different biochemical pathways. However, we can say that long-term storage is usually in organic form (seeds, stems of deciduous trees during winter). Some species, e.g., grasses, use both organic and inorganic storage forms. The latter might be considered as a more temporary storage, useful on sites where nitrogen availability fluctuates and does not always keep pace with a growing plant's need. In many temperate grasslands, the soil stores of easily available nitrate build up during autumn and become depleted during the season of peak growth in late spring. Leaf nitrate taken up early in the season and stored as such in the plant might then help to maintain growth longer than would be possible if the plant solely depended on uptake by roots. Leaf nitrogen storage and reduction may also be of advantage under desert conditions (Smirnoff and Stewart 1985), where the period when roots can grow and take up nitrogen may be very short.

As was said earlier, it can also be argued that nitrate reduction coupled to photosynthesis is a cheaper way to form proteins than dark reduction in the root (Gutschick 1981; Pate 1983). However, this is true mainly under light saturation and may be of limited value for shade plants (Smirnoff and Stewart 1985). Plants with CAM type photosynthesis (succulents) may reduce nitrate during daytime, and at light saturation, thus avoiding competition with the assimilation of carbon dioxide, which is absorbed from the outside air during the night, an advantage for a group of plants adapted to arid conditions (Smirnoff and Stewart 1985).

The form, ammonium or nitrate ions, in which nitrogen is taken up by a plant has a profound influence on its cation/anion balance, but it would be going too far here to discuss in detail the physiological and ecological implications of the ionic balance, and of the evolutionary adaptations of plants to environments with different ionic ratios. For further references, see e.g., Rorison (1980) and Pate (1980, 1983).

2.2 Nitrogen in Soils: Transformations, Uptake, Losses

2.2.1 Organic Matter Decomposition, Nitrogen Mineralization to Ammonium Ions, Microbial Immobilization and Root Uptake

As in plants, nitrogen in soil occurs both in organic and inorganic form. Organic nitrogen is in reduced form, some of it as amino and amide nitrogen, relatively easily available to decomposer organisms unless protected mechanically or chemically (see below). Another part of soil organic nitrogen occurs as a constituent of large and often resistant molecules with nitrogen in heterocyclic aromatic rings.

Inorganic nitrogen is usually fully reduced, ammonium, or fully oxidized, nitrate. Intermediary oxidation stages also exist but do not accumulate in measurable amounts, except for nitrite under special circumstances. There are transfers not only between soil and plant and between the various soil nitrogen pools, but also between the soil pools and the gaseous phase, where nitrogen compounds at different oxidation levels also occur (NH_3, N_2, N_2O, NO). Given all these possibilities, it is not surprising that the nitrogen chemistry and biochemistry of soils is complicated.

We have already stated that only a small part of the nitrogen store in the soil is available to plant roots at any given moment. Most is in organic form, usually in large molecules insoluble in water. Organic nitrogen in natural ecosystems originates from dead organisms, plants, animals, and microorganisms. The chemical bindings may, however, vary a great deal. Much of the nitrogen in fresh litter is still in protein form or in decomposition products of proteins, i.e., peptides and amino acids. These substances are attractive substrates for microorganisms, which often can use them as a source of carbon as well as of nitrogen. Their residence time in the soil is short, unless they are protected mechanically or chemically by association with less attractive substances in, e.g., cell walls. Presence of heavy metals may also inhibit decomposition (Rühling and Tyler 1973; Freedman and Hutchinson 1980).

The decomposition of litter does not mean that the litter nitrogen is immediately transferred to inorganic nitrogen or transformed into the limited number of low-molecular organic compounds in which it may be available to plant roots and mycorrhizal fungi. The chemical degradation of the litter is done by microorganisms, and even if they may produce extracellular enzymes, most of the nitrogen is taken up by the microorganisms themselves. The rate at which the microbial nitrogen is transferred to the available pool depends on the C/N ratio of the substrate and on the death-rate of the microorganisms. As many of the soil animals feed exclusively or preferentially on bacteria and fungal mycelium, the consumption by animals determines to a large extent the death-rates of the "primary" decomposers. Bacterial and fungal nitrogen is then concentrated before it is released with faeces and other metabolic products. As a rule, animals need a high intake of energy-rich substances (with high carbon content) for their respiratory activity. Nitrogen is used for the maintenance of their tissues and for propagation functions. Normally there is a much lower requirement for nitrogen than for energy-rich carbon compounds in the food of animals.

Animals specialized in food with high carbon/nitrogen ratios often live in symbiosis with microorganisms able to produce enzymes, e.g., cellulase, which facilitate uptake by the host of material otherwise protected by cell walls, at the same time as the cellulase makes the carbohydrates in the cell walls digestible. At the same time the microorganisms' own consumption lowers the carbon /nitrogen ratio. Such symbiosis has been described from the most different groups of animals, in ruminants and other mammals as well as in many insects and various components of the soil fauna. Wood-eating termites in tropical forests and savannas constitute a special case, as it has been reported that the guts of at least some species contain nitrogen-fixing organisms in addition to cellulase-producing ones (Breznak et al. 1973; cf. also Collins 1983). The rapid disappearance of woody litter by termite action is a well-known characteristic of tropical forests.

As far as nitrogen is concerned, the end product of the decomposition process as such is ammonium ions. Ammonium ions in water solutions are in equilibrium with undissociated ammonia molecules, but the amounts of ammonia are negligible until pH rises above seven. Such high pH values are seldom measured in forest soils (so-called rendzina soils with free calcium carbonate may have pH around 8). Yet similar high pH values may occur locally, in clumps of faeces or carcasses of animals of different sizes, or in decaying mushrooms. In such cases some ammonia may well be emitted to the atmosphere. In dense vegetation, e.g., under a forest canopy, much of that ammonia may be reabsorbed by the foliage and thus retained within the ecosystem. Yet under some circumstances foliage can release gaseous ammonia, probably mostly from senescent tissues (Farquhar et al. 1980, 1983; Lemon and van Houtte 1980).

The normal case, however, is that most of the ammonium liberated stays in the ecosystem, although rapidly removed from the soil solution along one of the following pathways: (1) uptake by plant roots (directly or via mycorrhizal hyphae), (2) uptake by microorganisms, (3) adsorption on the surface of soil colloids (in clay-rich soils partly followed by ammonium fixation in the lattice of certain clay minerals), and (4) chemical binding to organic substances. Any ammonium ions left in the soil solution may leave the soil with percolating water, but this is seldom an important pathway in natural ecosystems. Forests, heathlands, and grasslands are normally very tight systems with respect to ammonium nitrogen (as well as for organic nitrogen).

The different pathways just described need further comments. Root uptake and microorganism uptake are fairly straightforward processes, although it appears that microorganisms often have some competitional advantage over roots (Jansson 1958; Vitousek and Matson 1984). Turner (1977) has shown that stimulation of decomposers by addition of carbohydrate to a Douglas fir forest induced severe nitrogen deficiency in the trees and affected their internal nutrient distribution. On the other hand, conditions favouring nitrification tend to decrease the competitional advantage of the many microorganisms which prefer ammonium nitrogen to nitrate nitrogen.

Why so much of the plant uptake in natural ecosystems (and in many cultivated systems too) passes via mycorrhizal hyphae is still not satisfactorily explained, despite intensive research since the 1920s, when Melin (1925) started to grow mycorrhiza under sterile conditions in the laboratory. The most common form of mycorrhiza in coniferous forests, and also in many hardwood forests, is the ectomycorrhiza. Apparently this mycorrhiza formation offers advantages to both the tree and the fungal symbiont. The simplest explanation is that the tree's competitive ability for scarce nutrients (nitrogen, phosphorus, and possibly other elements) is improved by association with a microorganism, while the fungus has a secure supply of energy from the root (cf. Björkman 1943, 1970). While this "barter hypothesis" is still a reasonably good description of the mycorrhizal symbiosis, there is no full agreement on the mechanisms by which the root and the fungus establish their cooperation (Alexander 1983; Harley and Smith 1983; Nylund 1988).

Most experimental work on mycorrhizal physiology has been done either on excised roots or on mycorrhiza-infected plants growing in sterilized soil. It is therefore difficult to determine the real differences in uptake between mycorrhizal and non-mycorrhizal roots. In the case of excised roots, the main nutrient sink within the plant is removed; in the other case the true nutrient concentrations at the surface of the absorbing organs are unknown, and possibly different around organs of different geometry. However, Ingestad et al. (1986) have grown ectomycorrhizal and non-mycorrhizal seedlings of *Pinus sylvestris* in nutrient solution and found their relative growth rate and nitrogen productivity to be very similar, regulated by the nitrogen addition rate. There is thus reason to look for other explanations for the establishment of mycorrhiza than just the efficiency of the absorbing surface per se.

The fine roots of vascular plants are much thicker and therefore more energy-consuming structures than fungal hyphae. Even if the roots enlarge their surface by root hairs, it has been hypothesized that a comparable soil penetration may be achieved with less energy consumption by hyphae than by roots (Bowen and Smith 1981). Other possible explanations for the common occurrence of mycorrhiza in many ecosystomes may be sought among: (1) the active lifetime of both types of absorbing substances, (2) their vulnerability to various damaging influences, and (3) the way in which plant nutrients occur in the soil, which differs between most forest ecosystems, on the one hand, and arable land and many temperate grasslands, on the other. In the latter case, roots developing in spring find soil colloids loaded with relatively high concentrations of nutrients (originating from decomposing processes during winter, or from fertilizer). In forest ecosystems nutrients are more continuously released but seldom occur in high concentrations, at least not in the case of growth-limiting elements such as nitrogen and phosphorus. The dominant cation in non-fertilized soils is often calcium, and the observed ability of ectomycorrhizal hyphae to exude oxalic acid (as well as some other low-molecular organic acids) might facilitate the uptake of other cations, such as ammonium, potassium, and magnesium, which do not form virtually insoluble oxalates, as calcium does (Cromack et al. 1977, 1979; Sollins et al. 1981).

Several authors have suggested that mycorrhizal fungi may use low-molecular organic compounds. Now there seems to be a certain agreement (Alexander 1983; Bowen and Smith 1981; Nylund 1988) that the main nitrogen uptake by ectomycorrhizas is in inorganic form, where ammonium nitrogen is prefered. However, it is admitted that the geometry of the mycorrhizal system may give it better possibilities than roots to compete with microorganisms for, e.g., amino acids liberated by decomposition. In many ecosystems, such as grasslands and savannas, the most common mycorrhiza is the so-called endomycorrhiza of VA type (VA = vesicular-arbuscular, referring to the appearance of hyphae growing into the cells of the host plant; cf. Högberg 1986). This type of mycorrhiza differs in many respects from ectomycorrhiza (fungal species, growth mode, life cycle, host specificity), but apparently has a similar positive influence on the host plant, although even less seems to be known of the physiology of the symbiosis. Still further types of mycorrhiza are characteristic for special plant groups, such as orchids and ericacaeous species, where they are often necessary for the development of the host plant. In some cases, such as chlorophyll-lacking plants (e.g., the genus *Monotropa* and many juvenile and some adult orchids) there is evidence that the fungal symbiont apparently transmits to the host plant not only mineral nutrients, including nitrogen, but also organic substances, which originate from other green plants (Björkman 1960; Harley and Smith 1983). Read (1983) has produced evidence that ericaceous species, thanks to their mycorrhiza, may have better access to nitrogen bound in complex organic compounds than non-mycorrhizal specimens; at the same time the mycorrhiza offers protection against heavy metal poisoning.

Adsorption of ammonium ions to soil colloids is, as mentioned earlier, a removal from the pool of dissolved nutrients, but does not make them unavailable for plants; when roots or mycorrhizal hyphae deplete the soil solution of ammonium ions, such adsorbed ions will go into solution again according to well-known chemical principles. However, ion transport by diffusion is a slow process, so unless there is a mass flow of soil water, roots and hyphae have to grow close to the sites of adsorption. The energy cost for uptake from a soil increases in comparison with that from a nutrient solution. Lattice-fixed ammonium ions can also be redissolved, but this is a slow process of limited ecological importance under normal conditions and time perspectives (seasons, years, even decades).

Chemical binding of ammonium nitrogen in high-molecular organic substances in the soil is a very important and yet poorly understood process. Humus is the term for the soil organic matter which cannot macroscopically be recognized as plant or animal remains (Kononova 1975). Humus defined in this way always contains nitrogen, in lower concentrations in poor coniferous forest soils than in richer mull soils. The humus is very resistant to degradation, with half-lives varying from decades in some intensively cultivated organic soils to several thousand years for organic matter deep in mineral soil in certain soil types (as measured by radiocarbon dating). The chemical structure of humus is not well defined, even if fractions with different characteristics can be isolated by chemical methods (humic acids, fulvic acids). Much of the nitrogen appears to occur in heterocyclic aromatic rings, which together with the size of the molecules may account for the resistance to enzymatic degradation. Much of the carbon in the humus may originate from the lignin in plant cell walls, as terpenoid fragments can be obtained from both lignin and humus by chemical treatment. While many fungi and bacteria either lack lignin-degrading enzymes or produce them in small amounts, wood-degrading fungi of so-called white-rot type decompose lignin-rich plant residues relatively easily. Related soil-living fungal species can decompose at least part of the soil humus. When these fungi grow in fairy rings, their influence on the soil can be spectacular. The active growth zone of the fungus is often marked by luxuriant grass vegetation, surrounding a zone with lower yellowish vegetation, apparently suffering from lack of nutrients or otherwise unfavourable chemical or physical conditions. Despite such phenomena the concentration of lignin and other high-molecular polyphenolic compounds appears to be one of the important controlling factors for the rate of organic matter decomposition in the forest soil (Melillo et al. 1982; Berg 1986).

The formation of humus is a poorly understood process (Flaig et al. 1975). It has been suggested that at least some of the synthesis takes place in the digestive organs of soil animals, particularly earthworms, where both a suitable pH and relatively high concentrations of ammonia occur (Wittich 1952). This hypothesis would account for the higher nitrogen concentrations and the higher stability of mull humus as compared with that of mor humus, the former occurring in soils with many earthworms. Other differences in the biology between mull and mor will be discussed later, but it can be mentioned that mull (and to an even greater extent cultivated soils) can be looked upon as decomposition systems disturbed more or less continuously. In the case of mull at natural sites, the fungal and bacterial decomposers are continuously consumed by macrofauna and mesofauna, earthworms in particular, which also move material vertically and horizontally. The mor (or raw humus) represents a system in which the animal consumption takes place on a microscale. The many small animals in the mor consume both fungi and bacteria, but they do not move much material between horizons, and their action has little effect on the mechanical structure of the system. Therefore, a sampling for incubation experiments, for example, means a far greater impact on the mor than on the mull (Romell 1935, 1967), making conventional incubation tests of dubious value as assessments of mineralization in situ. They may still provide information on mineralization potential, but even then there may be interpretation difficulties when different humus layer types are compared (cf. Sect. 5.2.2).

Mull and mor, as just described, represent two contrasting types of soil horizons where a large part of the litter decomposition takes place. Yet other types of highly organic soil horizons, as well as transitional types, exist. The term *Moder* (originally a German word, while *mull* and *mor* are Scandinavian, introduced in the scientific language by P. E. Müller in the 1880s; see Müller 1887) is used for a humus layer with a less intimate mixture of mineral particles and organic matter than in the mull. Earthworm activity is low or absent, but large arthropods such as millipedes play an important role. Moder is sometimes considered as transitional between mull and mor, but it has characteristic traits of its own, and consequently deserves more attention. Simultaneous occurrence as well as true transitions are common between mull and moder, between moder and mor, and also between mull and mor. Simultaneous occurrence usually means that a highly organic horizon of either mor or moder overlies a horizon with a more or less well-developed mull structure (soil aggregates containing organic substances in intimate mixture with clay minerals).

Poorly drained soils also have their own characteristic humus types (e.g., German *Anmoor* and various peat forms), where permanent or temporary lack of oxygen affects both rates and pathways of various processes.

Many soils have been disturbed by human activities, both physically (plowing, drainage, etc.) and chemically (liming, fertilization). Natural disturbances also occur on a variety of scales, as will be discussed in later sections. All such disturbances may affect both soil morphology (e.g., humus form) and soil processes (e.g., decomposition and nitrogen turnover).

2.2.2 Nitrification

In many soils the mineralization of organic nitrogen is followed by further oxidation to nitrate ions or to gaseous forms of nitrogen (N_2O, NO). Most of these processes are mediated by bacteria. Some of them have been fairly well studied, while others are less well known, particularly with regard to their importance in an ecosystem context. Many nitrification studies concern the fate of added nitrogen fertilizers, which easily can be labelled with ^{15}N. Other studies concern the soil's potential for nitrification, investigated in model experiments. The natural rate of nitrification is most often estimated indirectly. For more detailed information than can be presented here, the reader is referred to recent reviews by Gundersen and Rasmussen (1988, 1990).

The rate of nitrification in a soil is affected directly and indirectly by many factors, such as temperature, moisture, C/N ratio (Kriebitzsch 1978), occurrence of inhibitors of the process itself, or of organic matter decomposition. Yet a prime prerequisite for nitrification is access to ammonium ions in the soil or, for some heterotrophic nitrifiers, easily available amino compounds. We have stated earlier that plant roots promptly absorb ammonium ions (as well as nitrate ions), while many microorganisms prefer the ammonium form. Some fungi cannot even use nitrate nitrogen. Concentrations of ammonium ions high enough to support an active population of bacteria using oxidation of ammonium to nitrite as their sole source of energy (e.g., the genus *Nitrosomonas*) only occur when the competition for nitrogen is low or moderate, i.e., when ammonia influx to the soil compartment (by ammonification or as input from outside) temporarily or permanently exceeds biological uptake. As will be discussed later, this situation is a common consequence of disturbances, reducing the biomass or its capacity to absorb nutrients. Similar conditions also occur naturally in ecosystems with climate-dependent variations in litter production, nutrient uptake, and soil processes. Examples can be found among deciduous forests and grasslands.

Access to substrate is, however, not sufficient to guarantee nitrifying activity in a soil. The best known nitrifiers are bacteria of the genera *Nitrosomonas*, which oxidize ammonium to nitrite, and *Nitrobacter*, which oxidize nitrite to nitrate. Both *Nitrosomonas* and *Nitrobacter* are favoured by alkaline to slightly acid soils and are unimportant in strongly acid environments. This does not necessarily exclude them from soils with an average acidity below pH 4.5 (Focht and Verstraete 1977). The heterogeneity of a soil means that there may be a large variation in many soil properties, including acidity, between microsites. In the case of pH, Nykvist and Skyllberg (1989) have found differences up to one-half unit between samples taken a few decimetres apart in a North Swedish mor layer. The variation in pH did not appear related to distance between samples or size of the sample plots but could be considered an intrinsic property of the mor layer. As their samples comprised several grams and a mor layer is a very heterogenic structure, it is quite possible that such differences in pH may occur also on smaller distances than decimetres. Yet pH is an important controlling factor, not only for the occurrence of nitrification, but also for any by-products that may be formed. As *Nitrobacter* seems to require somewhat higher pH than *Nitrosomonas*, some accumulation of nitrite may occur under certain circumstances. Gaseous products may also be formed, at different rates under different conditions.

It has been suggested (Verstraete 1981) that nitrification may represent a dominant global source of N_2O, and the production of NO may be important in some soils (Verstraete 1981), so the by-products of nitrification are interesting in connection with studies on global nitrogen cycling.

What has been said so far mainly concerns autotrophic nitrification, mediated by the reasonably well-known bacteria *Nitrosomonas* and *Nitrobacter* (together with a few other autotrophic bacteria). Nitrification also occurs in some soils too acid for known autotrophic nitrifiers, or lacking them for other reasons. It has been shown that nitrate formation may continue in the presence of inhibitors known to stop autotrophic nitrification (e.g., Kreitinger et al. 1985; further references in Gundersen and Rasmussen 1990). This indicates the occurrence of so-called heterotrophic nitrification, mediated by certain fungi (Focht and Verstraete 1977) or by methylotroph bacteria (Verstraete 1981). The importance and ecological preferences of these mostly facultative nitrifiers are still obscure, as are the by-products of their activities, although NO is suspected to be important under acid conditions. It is clear that heterotrophic nitrifiers form nitrate at a much slower rate than autotrophic nitrifiers (with the same biomass). However, a slow rate may be compensated for by a high biomass.

It remains to be stated that nitrification is an acidifying process. Under undisturbed conditions, when the nitrate formed is rapidly taken up by roots and reduced back to ammonium and other reduced forms, there is no net acidification. However, even in apparently healthy forest ecosystems, the two processes, nitrification and nitrate uptake, may be discoupled (Ulrich 1983), and a strong acidification of certain soil horizons may occur. If nitrate uptake lags behind nitrification more permanently, as would be the case at nitrogen saturation (see Chaps. 4, 5), the entire profile would be acidified. Such an acidification has been observed in long-term experiments with repeated additions of ammonium nitrate, as will be described in Sect. 4.4.

2.2.3 Denitrification: A Homeostatic Mechanism?

While the autotrophic nitrifiers use the energy released in the exothermic reactions involved in the transfer of nitrogen from a reduced to a more oxidised form, other bacteria exploit the oxygen in nitrate or nitrite in environments where oxygen supply is scarce or lacking. A stricter formulation would be that these organisms use nitrate or nitrite as electron acceptor instead of free oxygen molecules.

The number of bacteria with the ability to denitrify is large and diverse. Many use organic compounds as an energy source, while others may get their energy from reduction of hydrogen or reduced sulphur compounds. Not all possess all the enzymes necessary to reduce nitrate all the way to molecular nitrogen; in some cases N_2O is the main product, while others start from nitrite rather than nitrate. Low pH is usually an inhibitory factor, in particular for the last stage, from N_2O to N_2 (Nömmik 1956; Wijler and Delwiche 1954; Klemedtson and Svensson 1988).

The main controlling factor for denitrification, besides the occurrence of nitrate or nitrite, is the oxygen supply. The formation of nitrogen oxide-reducing enzymes is repressed by oxygen, – a natural reaction, if we look upon the use of nitrogen compounds as substitute electron acceptors under conditions where the normal acceptor, oxygen, is absent or scarce. However, there are bacteria which can use O_2 and NO_3 simultaneously as electron acceptors. Conditions suitable for denitrification often exist in the transitional zone between a well-aerated topsoil and a waterlogged or otherwise anaerobic subsoil. Yet the amount of nitrate and nitrite is soon exhausted in the anaerobic zone, and new production takes place only under aerobic conditions. Transport by diffusion is a slow process for solutes in the soil, in contrast to gas diffusion in aerated soils. High denitrification rates are therefore expected (and measured) in soils with repeated cycles of waterlogging and drying out rather than in the permanent transitional zones described above, although they may possess high potential for denitrification. There may also be large differences in redox conditions within and between soil aggregates; although diffusion distances are small, denitrifying microsites may occur frequently inside biologically active aggregates even in soils with good aeration.

In the previous section we discussed a discoupling between nitrification and nitrogen uptake by plants, with its consequences for soil acidification. The relations just described between nitrification and denitrification also imply the possibility of a discoupling of the two processes, at least at a microscale in the soil (e.g., within and between soil aggregates). Theoretically a discoupling might also be possible at larger distances, if soil water is moving from a nitrification zone to a denitrification area, e.g., from an infiltration area to an exfiltration area (Davidson and Swank 1986). This possibility ought to be considered when mass balances for catchments are used for estimates of nitrogen inputs and outputs, but at present it is difficult to assess the importance of this discoupling in quantitative terms.

Denitrification can also occur as a purely chemical process. Nitrite, which is far more reactive than nitrate, can react with ammonium or amide groups to form molecular nitrogen. Ferrous iron can also react with nitrate.

From an ecological viewpoint the denitrification process can be looked upon as a homeostatic mechanism preventing a destabilization of the ecosystem. It is often tempting to apply an anthropomorphistic way of looking at an ecosystem as a kind of seemingly intelligent superorganism. Such a view could generate counterproductive hypotheses and conclusions. Yet it is absolutely necessary to study all kinds of stabilizing factors and negative as well as positive feedbacks operating in ecosystems. It has been suggested (Lovelock 1979, *The Gaia Hypothesis*) that the biosphere as a whole has a considerable ability for self regulation, and that nitrogen transformations play an important part in that regulatory system.

In some cases homeostatic mechanisms are weak and of limited influence on energy and nutrient flow, while they still determine species distribution. Many sea shores provide examples where gains and losses of both material and energy depend on extremely variable physical factors, yet the ecosystem remains relatively constant in species composition.

Forest ecosystems are very different from such a situation. Many of the transfer processes in a forest may be described as interactions within the system, between the tree canopy, lesser vegetation, and soils (including soil organisms). Fluxes between pools are reasonably predictable, and even if we do not know all the feedback mechanisms, we can be sure that many of them contribute to the apparent stability of the forest ecosystem from year to year. Yet part of this persistence may be inertia rather than stability in the true sense, depending on the ability of the tree stand, with its large biomass and nutrient stores, to buffer at least moderate changes in environmental factors. If the tree stand is removed by nature (windfelling, fire, insect attacks) or man (clearfelling), the new ecosystem may or may not develop in the same way as the previous one.

Most natural terrestrial ecosystems accumulate nitrogen, at least in some stages of their development. An uninterrupted accumulation, unless it is accompanied by a permanent immobilization, would lead to a situation where nitrogen would occur in excess. Yet nature has some safety valves. There are at least five ways in which the increase in the pool of recycling (and thus the easily available) nitrogen can be counteracted:

1. permanent immobilization in, e.g., peat or sediments,
2. leaching,
3. denitrification,
4. volatilization in fire, and
5. harvest or other organic matter removal.

Peat formation occurs only in poorly drained sites, mostly in humid climates. Removal of organic matter is characteristic for managed systems, but can be important in certain types of natural sites in the form of grazing or erosion. The mechanisms responsible for ecosystem losses of nitrogen in most types of natural terrestrial ecosystems must be sought among the other three ways. All play roles in ecosystem nitrogen budgets, but with strong variation in relative importance. We shall come back later to the possible role of denitrification as a safety valve in certain ecosystems.

We stated earlier that nitrification is an acidifying process. In general, denitrification and nitrogen reduction counteract acidification. Hydrogen ions are consumed in the process. Yet it remains uncertain how significant this ecological process is in terrestrial ecosystems at present. Acidified ecosystems, where the process should be important, often have depressed or intermittent nitrification. In such cases denitrification may be expected to lag behind nitrification, without ability to fully counteract the acid pushes described by Ulrich (1983). A regular denitrification of the nitrate component of the acid deposition would mitigate the soil acidification, but there is little evidence that this is an important process. If the present trend in atmospheric pollution continues, with increasing NO_x and decreasing sulphur emissions, denitrification might become a more important deacidifying process also in terrestrial ecosystems (with N_2O and possibly NO as by-products).

Denitrification may be more important in lake and some wetland systems, where nitrogen often occurs as nitrate, although in low concentrations; here favorable conditions for denitrification often prevail in the surface layer of the sediments (Shindler 1986).

2.2.4 Nitrogen Fixation

Biological fixation of atmospheric nitrogen by free-living bacteria or such living in symbiosis with species of legumes (*Rhizobium*) is probably the most studied of all the processes in the nitrogen cycle. This does not mean that our knowledge is satisfactory in all respects, but we can here restrict ourselves to a few relevant statements concerning the possible effects of changes in chemical climate or management on the nitrogen cycling.

In a global nitrogen budget, biological fixation represents the largest input for terrestrial ecosystems (Fig. 1.1). The input is, however, very unevenly distributed. Ecosystems containing plants in symbiosis with nitrogen fixers, such as alder forests, *Myrica* wetlands, *Acacia* savannas, or legume croplands, may show fixation rates between several tens and some hundreds of kilograms per hectare and year. Free-living bacteria are usually less effective than the symbiotic ones, but in many ecosystems they contribute more nitrogen than that added with wet and dry deposition in unpolluted areas. Nitrogen fertilization depresses fixation of both symbiotic and free-living bacteria, so it might be assumed that emissions of NO_x and NH_4, if intensive, might also affect fixation. Other factors of importance for the rate of fixation are soil pH (most nitrogen fixers prefer relatively high pH, even if some may be active down to pH = 4.5 or lower), and the supply of nutrients other than nitrogen. Quite a few plants associated with the nitrogen-fixing bacteria are relatively demanding with respect to mineral nutrients (Pate 1986), and it has been speculated that the sensitivity of the process to low pH has something to do with the low solubility of molybdenum in acid soils. Molybdenum is a constituent of one of the nitrogenase enzymes and essential for all biological nitrogen fixation.

As with all adaptations, symbiosis with nitrogen-fixing bacteria has an energy cost, which reduces the competitive advantage of access to molecular nitrogen (Pate 1986). Consequently, plants with nitrogen-fixing symbionts are common only in sites where available nitrogen is scarce. This is mostly the case in early stages of primary succession and also in some types of secondary succession, where plant species with symbiotic fixation may be abundant (*Alnus*, *Hippophae*, some legumes). Lack of nitrogen-fixing species and low rates of non-symbiotic fixation are common in later successional stages, but epiphytic lichens both in tropical rain forests and in old stands of Douglas fir may have active symbiotic fixation (Waring and Schlesinger 1985).

Thus, there are many different ways in which the nitrogen fixation rate in an ecosystem can be affected by humans. So far most of the impact has been due to agricultural land use and forest management. Yet the present scale of industrial emission of nitrogen compounds is bound to have effects on nitrogen fixation in the environment: locally, regionally, and even globally. The interaction between the various substances in polluted air (such as organic substances and acidifying sulphur compounds) with nitrogen compounds further complicates the picture.

2.3 Nitrogen and Transport Mechanisms in Terrestrial Ecosystems

2.3.1 Within-Site Transfers

Every textbook on ecology contains one or more diagrammatic pictures of the nutrient cycle within the ecosystem. Figure 1.1 is such an attempt to illustrate the global nitrogen cycle. We have already discussed several of the processes

Table 2.1. Transport processes within terrestrial ecosystems with special significance for the internal cycling of nitrogen and other elements. The degree of biotic and abiotic control noted concerns direct control only and is to some extent arbitrary

Process	Compartments concerned		Controlling factor[a]	
	Source	Sink	Biotic	Abiotic
1. Plant uptake (including fungal uptake)	Soil, different horizons	Plant	+ + +	–
2. Litter fall (above-ground)	Above-ground vegetation	Ground surface	+ + +	+
3. Redistribution in plants	Plant	Plant	+ + +	–
4. Canopy leaching, stem flow	Plant	Soil	+ +	+ +
5. Soil leaching	Upper soil horizon	Lower soil horizon	+ +	+ + +
6. Soil capillary rise	Lower soil horizon	Upper soil horizon	–	+ + +
7. Litter incorporation in mineral soil	Litter layer	Mineral horizons	+ + +	–

[a] The plus signs signify degree of control: + some, + + moderate, + + + strong. Minus sign: no direct influence.

concerned in Section 2.2. The central theme in this section is not so much the transfer processes per se, but rather how transport mechanisms in the ecosystem, both for nitrogen and for other elements, are influenced by changes in nitrogen abundance.

Table 2.1 lists some of the processes which are of special significance for the local nitrogen cycle, either because of the quantities transferred, or because the nitrogen transfer is rate determining for other important ecosystem processes.

Like most classifications, Table 2.1 presents a highly schematic picture of a complex reality. Processes designated as biologically controlled are of course also influenced by the abiotic environment, both directly and indirectly. In the case of soil leaching, the direction and intensity of flow is mainly abiotically regulated, while biological processes influence the concentrations of solutes in the soil solution. The influence of animals has been left out, except for process 7, where larger soil animals (earthworms, ants, rodents, etc.) are the main agents. Indirectly, animals influence all processes designated as biologically controlled, by influencing the growth and vigour of plant components of the ecosystem. More directly, sucking and leaf-eating insects affect canopy leaching, by exposing cell components otherwise protected by living membranes, cuticle or bark to rainwater, before or after passing through the guts of the insects.

It should perhaps be pointed out (see Table 2.1) that while plant nutrients are transferred from soils to green plants by a single, biologically controlled process, uptake by roots (and mycorrhizas) (1), the return from plant to soil is more complicated, by litterfall (2), canopy leaching (4), and excretions from roots and root abscission (a process which is not a transport in the strict sense, even if it means a transfer from a living to a dead compartment).

The quantitatively most important transport processes in the case of nitrogen are (1) the uptake by roots, (2) the loss by litterfall, and (3) the redistribution to sinks within the plant, such as photosynthesizing organs during the vegetation period, growing organs (including roots) during their active period, and to storage organs, which in evergreen conifers include the needles. The main flow direction in the redistribution process is thus upward, from roots in various soil horizons to other organs, in forests particularly to above-ground parts of the trees. This is the main mechanism for ecosystem retention of both mineral nitrogen and easily movable cations, which we have mentioned several times.

The role of the decomposer fungi in the retention of nutrients not only consists in immobilization in their tissue, as also done by other soil organisms. In a litter layer with a high C/N ratio, the litter is often enriched with nitrogen (and phosphorus), also in absolute amounts, during the early decomposition stages. The increase runs parallel with fungal ingrowth in the litter (Berg and Staaf 1981), which suggests that the fungal hyphae transport nutrients from below, where nitrogen is released at the same rate as mass loss (Berg 1986).

Most of the nitrogen taken up and stored in above-ground parts of the vegetation is returned by litterfall (2), where the total amount of litter over a period of years is well related to biological production. Yet the seasonal and annual variation in litterfall is often highly dependent on climatic factors. For some nutrients, potassium in particular, canopy leaching (4) is a pathway of at least similar importance as the litterfall for the return to the soils (Parker 1983). In the case of nitrogen there are usually low concentrations of mineral nitrogen in the throughfall on unpolluted sites, usually lower than in the incoming rain. Yet there is usually some organic nitrogen in the throughfall water, and as some of this may be easily decomposed amino acids, this transport could have some importance for the nutrition of some ecosystem components, epiphytes and soil-living mosses and lichens. The dependence of feather mosses on throughfall nutrients has been shown (Tamm 1953).

The throughfall water reaching the ground surface has, according to the above, already a load of dissolved substances. Normally most of them are leached from the canopy (4), but some originate from other sources. Nitrogenous compounds can be both leached and absorbed by the foliage, depending on pollution level, nitrogen status of the tree, and occurrence of leaf-eating organisms. As nitrate ions have weak affinity to absorbing surfaces such as bark and cuticle, nitrate is absorbed to a lesser extent than ammonium ions, and amounts of nitrate in throughfall can therefore be used as an approximate measure of dry deposited nitrate in polluted areas but not on very clean sites (Grennfelt and Hultberg 1986).

The percolating water gets further additions of solutes on the way through the litter layer (5), where decomposer organisms are active but uptake by roots and mycorrhiza is still absent or limited. Further down there is a competition between release and removal of soluble substances. In a normal soil profile well penetrated by roots this results in low concentrations of ions in the water leaving the profile.

Also much of the organic matter in the percolating soil water (partly as colloids, partly in true solution) disappears from the solution on the way down. This has an explanation other than the removal of nutrient ions. In many profiles there is a gradient in pH with depth, and it has long been assumed that some organic and inorganic colloids are precipitated when reaching a level with higher pH than in upper horizons. While this may be the actual chemical mechanism, many authors believe that microbiological decomposition of the organic complexes during the transport plays a large role in their disappearance, and probably also for the maintenance of a pH gradient (Tamm and Hallbäcken 1988). As acid groups (carboxylic or phenolic) form the most reactive parts of the organic substances, they will be attacked first during degradation. If microbial decomposition plays this important role, most of the nitrogen in the organic matter would be expected to be incorporated in the cells of the microorganisms (especially in soils with a high carbon/nitrogen ratio) or taken up by roots, if there is a good root penetration.

It is obvious that nutrient transport with soil water is a process which also can proceed from below (6). Yet there is only a limited range of soils which have pores large enough for fast water movement and yet admit a capillary rise from a ground water table below the rooting zone. The process is, however, of extremely great importance in arid and subarid regions. In dry climates the risk of salinization is high, a soon as the water table is high, due to either natural reasons or irrigation.

Process 7 in Table 2.1 is also, in principle, a two-way process. The classical example is Darwin's earthworm studies (1881), where he was able to show that the animals drag litter into the mineral soil as well as depose material from below on top of the soil surface. As already stated (Sect. 2.2.1), there are large ecological and microbiological differences between soils with an active biological mixing of litter material with mineral soil (type case: mull) and those where this mixing is lacking. As discussed earlier, we have found differences in nitrogen turnover between these two types. We shall not comment further on these differences here, but merely point out that differences in transport systems in the soil are involved and that some of the soil characteristics depend on the occurrence of biological transport agents.

While there is an increasing recognition of the role of living organisms, plants in particular, in the conservation of nutrients within the ecosystem, the term normally used for the end result of the biological processes in Table 2.1 is the relatively unspecific term biological retention (Bormann and Likens 1979). Perhaps it should be remembered that Russian pedologists have long recognized what they call the "sod process", primarily meaning the incorporation of soil organic matter in the upper mineral soil, but also covering the pumping action of the roots, maintaining the nutrient store in the top soil despite the downward movement of the soil solution.

2.3.2 Inputs to and Outputs from Terrestrial Ecosystems

Most natural terrestrial ecosystems are characterized by a relatively closed nitrogen cycle, i.e., inputs to the system and outputs from it are small in relation to the amounts cycling between soil and vegetation. Measured outputs in the runoff from catchments are often smaller than the estimated input by wet and dry deposition and this is particularly true at low deposition levels (Abrahamsen 1980; Grennfelt and Hultberg 1986).

Published nitrogen budgets for catchments and ecosystems are seldom complete, as the only easily measured budget posts are wet deposition and nitrogen in runoff. Dry deposition and nitrogen fixation are usually calculated, the former from physical models and the latter at best from experimental results scaled up from single plants or soil samples. Denitrification is even more difficult, as the measurements done usually concern denitrification potential rather than actual rate.

Yet there are many ecosystems where we know that nitrification is of little importance, and as denitrification requires nitrate as substrate, this process would also be unimportant, the only substrate available being the limited amounts deposited from the atmosphere. In such cases we can be confident that nitrogen is accumulating in the system. Such an accumulation has also been measured during succession in an aggrading forest ecosystem (Bormann and Likens 1979). In primary succession as well as in some secondary ones much of the early accumulation can be ascribed to symbiotic fixation (Crocker and Major 1955). Yet there is good evidence from both catchment budgets and soil studies that at least forest ecosystems continue to accumulate nitrogen even after an initial phase of intensive fixation. Evidently such an accumulation cannot go on forever, and in natural ecosystems there are corrective mechanisms, causing losses of nitrogen, usually by nitrate leaching (with or without accompanying denitrification). Natural agents such as storms, insect defoliation, and, above all, fire may destroy the existing vegetation and stimulate both nitrogen mineralization and nitrification, leading to temporary losses of nitrate, as will be further discussed in Chapter 3. A schematic picture of the nitrogen cycling in an undisturbed forest is presented in Fig. 2.6 (upper part). The lower part of Fig. 2.4 shows the various possibilities for increased losses of nitrogen after disturbances, natural or anthropogenic.

The result of all these processes is the existing landscape, where within most biomes there appears to be a correlation between forest productivity and the nutrient cycling (Cole and Rapp 1981). While the relationship carbon store/net primary production appears straightforward, or possibly a question of the hen or the eggs, the nitrogen store/net primary production relationship is only partially explained by the fact that nitrogen availability often limits the production. How were the higher nitrogen stores of the better sites created from the

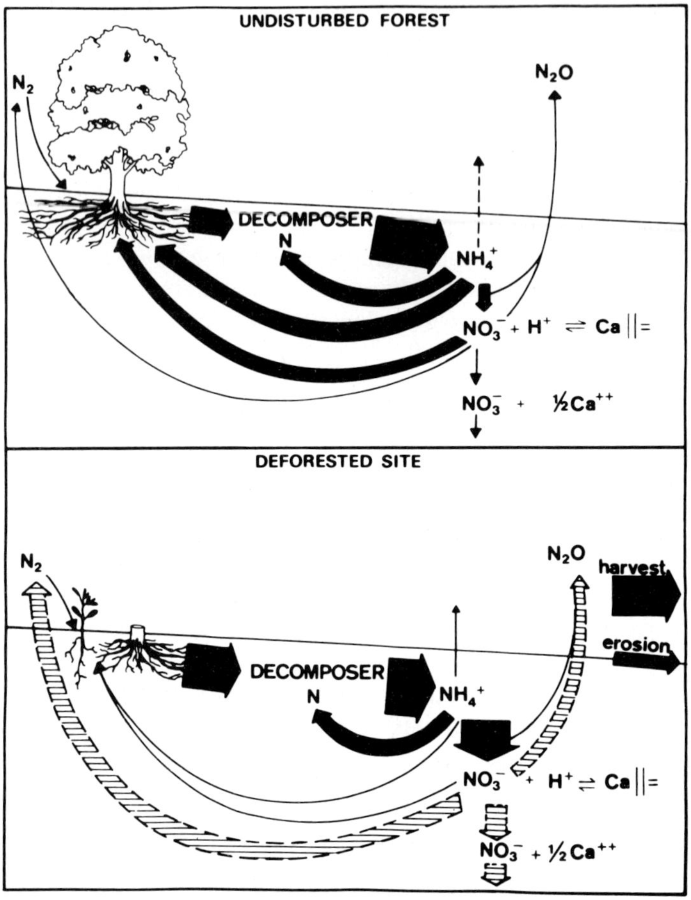

Fig. 2.6. The nitrogen cycle in an undisturbed forest and a deforested site. The system represented is a relatively fertile site before and 2–3 years after deforestation. *Dashed arrows* represent possible alternative pathways for losses of nitrogen from the site (Vitousek 1983)

beginning? There are three main possibilities: (1) Better conditions for nitrogen fixation, at present or earlier in site history, (2) fewer losses due to fire or other disturbances, and (3) transport of nitrogen from other ecosystems.

We have reasons to believe that all three possibilities are realized in nature.

Indirect evidence for the first mechanism can be obtained from the relation between the nitrogen and carbon stores (and also the C/N ratio) on the one hand, and base saturation (Dahl et al. 1967, see also Sect. 3.3) on the other. As most nitrogen-fixing bacteria (and also many of the plant species with symbiotic fixation) prefer soils with higher pH and more calcium ions, and such soils usually have higher nitrogen stores and lower C/N ratios than more acid and base-deficient soils, there are good reasons to believe in a causal relationship. It should also be remembered that soil humus formed at higher pH is more stable and contains higher nitrogen concentrations than acid mor humus with high proportions of fulvic acids. Formation of more stable humus might be a mechanism for nitrogen retention, especially if the humus also is richer in nitrogen.

A case where mechanism (2), differences in losses of nitrogen, appears to have had a decisive influence is discussed in Section 3.4 and illustrated in Fig. 3.3. The third mechanism, transfers from other ecosystems, deserves special mention, as quantitative documentation of such cases is rare in relation to the importance of this ecological process. Natural transfers (which of course are two-way processes, as will be further discussed in the following chapters) can occur through animals or wind (e.g., accumulation of wind-blown litter), but the most important instance is transport by water. The fertility of alluvial soils was known in ancient Egypt and Mesopotamia, and it was recognized that the river-borne silt contributed to it. Highly productive natural ecosystems can often be found on valley bottoms, where occasional floodings deposit silt. Examples are poplar forests in various parts of the world and the valley-bottom redwood (*Sequoia sempervirens*) forests in California. In such cases, most of the suspended and dissolved material deposited on the valley bottom has been low in nitrogen, but not free of it. Part of the positive effect might be due to improved conditions for nitrogen fixation, rather than the transport of nitrogen.

A case where transport of organic nitrogen in water must have played the major role has been described by Holmbäck and Malmström (1947) from northernmost Sweden, where water from an oligotrophic wetland, rich in dissolved and suspended organic matter, was used to temporarily irrigate a forest site, which had been extremely unproductive due to lack of available nitrogen. The response in tree growth was dramatic and lasted for several years after the end of the irrigation period (Fig. 2.7).

The most common case of nutrient transport by water from one site to another is that of downward slope. In hilly country there is normally a gradient in site quality along a slope, with the lowest qualities on ridges and the highest on lower slopes. Where exceptions to this general rule occur, they can usually be explained by counteracting adverse factors such as waterlogging or high frost frequency near the valley bottom. The possibility for transport of water and solutes along a slope, visually estimated from the topography, is used in the Swedish system for forest site evaluations from site factors (Hägglund and Lundmark 1977), where it makes a statistically significant contribution to the accuracy. In southeastern Ohio, Carmean (1967) found the site index for black oak to decrease with slope steepness in slopes facing south and west, but in slopes facing north-east the index increased with steepness, so the best sites were found on steep slopes. The first mentioned decrease in the site index could be explained as a drought effect, but the simplest explanation of the good growth in the north-east-facing slopes seems to be nutrient flush along the slopes.

It can be argued that most of the dissolved material in the subsurface flow is low in nitrogen, but plant roots and microorganisms are very efficient in us-

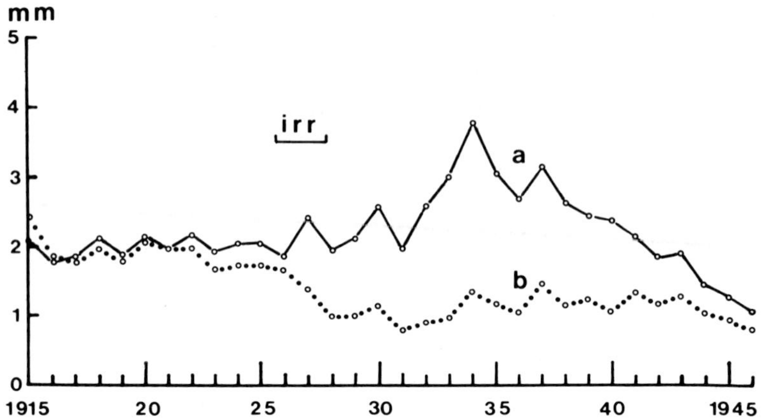

Fig. 2.7. Average widths of annual rings in pine 50 to 60 years old, Ruuttirova, Swedish Lappland. The *solid line* represents the mean of 15 trees growing in an area subject to irrigation from 1926 to 1928 (*irr*) and the *broken line* represents the mean of five trees growing on nearby unirrigated ground. The diagram illustrates one of several similar comparisons (irrigation during different periods) (Holmbäck and Malmström 1947)

ing nutrients occurring in low concentrations, especially if the nutrient solution is regularly renewed. If some of the flow passes over the soil's surface or through the A horizon, it may contain considerable amounts of organic matter, always with some organic-bound nitrogen. The existence of a surface or a subsurface flow along a slope does not guarantee that the ecosystem concerned has a net accumulation of nutrients in a particular year. It is difficult enough to measure the amounts of nutrients in the flow (cf., however, Troedsson 1952, for data on base cations), so measuring differences in inflow and outflow for a segment of a slope would be extremely difficult. Yet we have already mentioned the changes between accumulating and degrading phases in forest ecosystems (Bormann and Likens 1979), and in the course of time, downslope ecosystems might get repeated changes to accumulate nitrate trickling down from degrading upslope sites. As nitrate ions are always accompanied by cations (cf. Fig. 2.6), periods of intensive nitrate leaching also mean periods of increased cation transport. One might speculate whether the periods between degrading phases, characterized by low but steady downslope flows of cations accompanied by anions such as bicarbonate, sulphate and chloride, will allow downslope sites to build up larger pools of exchangeable cations than, e.g., ridge sites or well-drained valley bottom sites, where the only sources of cations are weathering and atmospheric deposition. An excellent description of the chemical differences between flushed and other sites in the English Lake District has been made by Gorham (1953).

It is a disturbing fact that we have so little hard information on the long-term nutrient dynamics of sloping sites. Considering the accumulating nature of most forest ecosystems, even rather small annual additions of nutrients may be enough to create the often strong site differences that we can observe, also under circumstances where the better water supply downslope appears to play a minor role. Most ecosystem studies concern either sites with predominantly vertical water movement, or whole catchments with their mosaic of sites. The situation is quite different in aquatic studies, since Odum's (1957) classical study of spring ecosystems, and also in wetland ecology there has been much discussion of nutrient transport with flowing water (Heinselman 1970).

3 Nitrogen-Limited and Nitrogen-Depleted Terrestrial Ecosystems: Ecological Characteristics

3.1 Definitions

Nitrogen-limited ecosystems are those where primary production increases markedly if nitrogen fertilizer is supplied. Nitrogen-depleted ecosystems are those where natural or human influences have led to large losses of nitrogen from the site. Nitrogen-depleted ecosystems are usually nitrogen-limited, while the reverse may or may not be true.

Ecosystems with low amounts of nitrogen cycling between soil and vegetation are common in all parts of the world. There can be different reasons for such situations. Some other factor, such as water supply, may set narrow limits to plant growth and nutrient uptake, and in this case the ecosystem is not nitrogen-limited and it is usually not depleted of nitrogen, as there are no amounts of importance which can be removed. We shall not discuss this further except to mention that severe lack of certain plant nutrients on a site (in particular phosphorus) may interfere both with demand (plant uptake) and supply (release of mineral nitrogen by organic matter decomposition). Plants may then show symptoms of deficiency in nitrogen as well as of phosphorus, but phosphorus addition may also alleviate the nitrogen deficiency, while addition of nitrogen aggravates the deficiency in phosphorus. This situation is not uncommon in sites low in available phosphorus, e.g., drained peatlands. Vitousek (pers.comm.) has coined the expression "phosphorus limitation in disguise" for the situation in which plants on low-phosphorus sites exhibit symptoms of nitrogen deficiency.

3.2 Nitrogen in Forest Successions

In cases where the supply of nitrogen to plant roots restricts growth and photosynthesis of the plants to a larger or smaller extent, one reason for the restricted supply may be a low total content of nitrogen on the site. This is the case in early stages of primary successions, as a recently exposed mineral soil is virtually devoid of nitrogen (Vitousek and Walker 1987). The situation favours plants with symbiotic nitrogen-fixation, and they often dominate early stages of primary successions (alders colonizing upon glacier retreat (Crocker and Major 1955), alder and *Hippophae* on the slowly rising shores along the Bothnian Gulf, etc.). Thanks to the nitrogen fixation (including non-symbiotic fixation) and the small atmospheric inputs of bound nitrogen, the amount of nitrogen in the ecosystem increases, as does the pool of available nitrogen, at least for some time. The nitrogen fixers then lose their competitive advantage.

Late stages in a forest succession do not have many nitrogen fixing organisms. Exceptions to this rule are found among epiphytes with their associated microorganisms, which lack access to soil nitrogen. Also, soil-living lichens, which normally obtain most of their nutrients from rain and throughfall, may fix nitrogen, usually in rather limited quantities. It is clear that the rate at which a nitrogen pool is acquired during a succession differs with climate, soil material, colonizing plant species, and other factors influencing the system. Data for nitrogen fixation over a forest rotation period are scarce, for natural reasons, but Bormann and Likens (1979) estimate for Hubbard Brook sites that 70% of the nitrogen store is derived from fixation and 30% from deposition. Generally, the total store of nitrogen increases over a succession period, but this does not necessarily apply to available nitrogen. In areas with an unfavourable climate and where the dominating plants produce slowly decomposing litter, the available nitrogen may well decrease in late successional stages, due to immobilization (Hesselman 1937; Sirén 1955).

Cole et al. (1967) calculated that in a second-growth Douglas fir forest there was an annual accumulation of nitrogen not only in trees (23.6 kg ha^{-1}) but also in the litter layer (11.5 kg), while the mineral soil lost almost the same amount (34.6 kg). A redistribution of this kind is often assumed to take place when natural hardwood forests are replaced by conifer plantations. The stronger leaching of nitrate from spruce plots (planted after beech) than from adjacent beech forest in Solling (Matzner 1988) may also be taken as evidence of degrading processes in the mineral soil.

According to Miller (1981), the demand for nitrogen by a forest stand peaks when the canopy is about to close (Fig. 3.1), and consequently forest fertilization should aim at rapid canopy closure. After that period, rapid nutrient cycling between canopy and soil is established, and no dramatic effects of fertilization are to be expected. Many results from plantation forestry, particularly in Australia and New Zealand, support Miller's theory, but positive fertilizer responses, let alone that they are more moderate than in some younger stands, have been obtained also in relatively dense stands in different parts of the world, e.g., in Germany (Kenk and Fischer 1988), in Japan, and in the Pacific

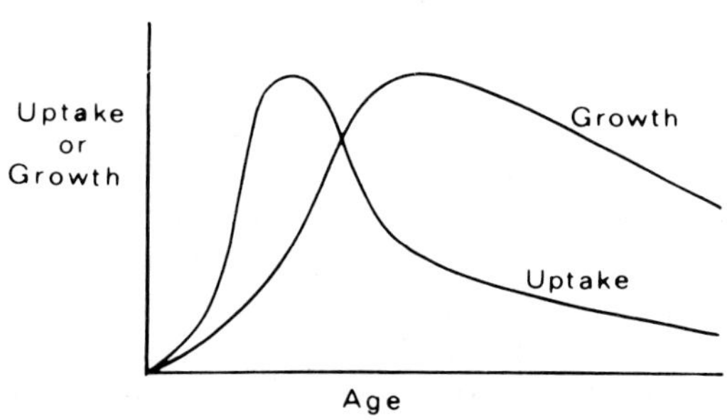

Fig. 3.1. Generalized pattern of nitrogen uptake and stem growth with age in an even-aged tree crop (Miller and Miller 1988)

North-West of North America (Stegemoeller and Chappell 1989). On the basis of a broad survey of data linking nitrogen cycling to ecosystem productivity, Melillo and Gosz (1983) conclude that increased atmospheric deposition of nitrogen would increase global forest production, and that much of that increase would take place in the temperate region. In the boreal forest, which typically is relatively open, positive responses to nitrogen are common in all age stages, even if fertilization practice avoids nitrogen addition to very young stands, where competing vegetation usually benefits more than the tree saplings from the treatment. Cole and Rapp (1981) conclude on the basis of nutrient cycling studies during the International Biological programme (IBP) that boreal tree species are more efficient than temperate species in producing biomass per unit nitrogen, apparently because sites there are more often nitrogen-limited. According to Vitousek (1982), boreal tree species should then have a high nitrogen use efficiency.

Very little experimental work has so far been done on limiting nutrients in undisturbed humid tropical forests, while it has been known for some time that plantations in tropical and subtropical regions may respond to fertilization with phosphorus and also nitrogen (e.g., Watson 1973). Jordan's (1985) conclusion that "the productivity of native (tropical) forest is seldom seriously inhibited by lack of phosphorus" is based on the observation that tropical forest ecosystems have biological mechanisms (i.a., mycorrhizas) which keep phosphorus cycling in a plant-available form. However, as strongly phosphorus-fixing soils are widespread in the tropics, experimental tests of the conclusion are desirable. In analogy with the behaviour of nitrogen-limited ecosystems, it seems conceivable that an otherwise undisturbed phosphorus-limited ecosystem adapts its primary productivity to the amount of phosphorus which can be kept in cycling by the ecosystem. Visual symptoms of deficiency in phosphorus, observed in agriculture and plantation forestry, do not appear to the same extent, if the plants are given time to adapt to a low nutrient flux density (Ingestad 1987). On the one hand, there seems to be evidence for a relatively satisfactory supply of nitrogen in most lowland tropical rainforests, where symbiotic nitrogen fixation is widespread and the carbon/nitrogen ratio of the litter is often low, suggesting low nitrogen use efficiency (Vitousek 1984). On the other hand, phosphorus use efficiency is high (Vitousek 1984; Melillo and Gosz 1983). Mountain rain forests in the tropics often have lower leaf concentrations of nitrogen and higher nitrogen retention than lowland forests, which suggests nitrogen limitation (Vitousek and Sanford 1986).

As mentioned in Section 2.1, effective redistribution of nutrients within the plant is an important competitive advantage for species occurring in poor environments. It is thus not a coincidence that species characteristic of late successional stages in environments low in available nitrogen have very efficient redistribution mechanisms. In the boreal forest this applies to conifer trees and ericaceous dwarf shrubs. Cole and Rapp (1981) consider evergreen conifers as having less efficient internal cycling of nitrogen than deciduous species, but on the other hand they produce more biomass per unit nitrogen. As much of the nutrient translocation in evergreen conifers takes place from older needles on a parent shoot to the growing daughter shoots and their needles (Sect. 2.1) during the elongation period and shortly afterwards, figures based upon annual budgets may not reveal the whole truth. A consequence of an effective redistribution is a litter low in plant nutrients, including nitrogen, and therefore prone to immobilize nitrogen and some other nutrients rather than release them in the early stages of decomposition (Berg and Staaf 1981).

The immobilization of nitrogen in peat on poorly drained sites represents a case with a more permanent sink for nitrogen. A sort of steady state may then be created, with small vegetation changes, an endpoint in the long-term succession of the wetland, apparently stable as long as external conditions, including the supply of nitrogen and other nutrients, remains constant. Small (1972) found very efficient redistribution of nitrogen in evergreen bog plants.

3.3 Biological Control of Nitrogen Cycling

The losses of nitrogen from undisturbed forest ecosystems are usually lower than the additions (Abrahamsen 1980; Grennfelt and Hultberg 1986). Living vegetation exerts a control over the leaching losses as long as the amounts of essential nutrients are not far in excess of plant demand (Bormann and Likens 1979; Gorham et al. 1979). This control is particularly important in the case of nitrate, which in most soils is not absorbed on soil colloids, and then acts as a vehicle for cation leaching, as a leached nitrate ion is always accompanied by a cation (metal cation or hydrogen ion). But potassium and other metals are also enriched by roots and pumped back to plants from different soil horizons, eventually being deposited on top of the soil with litterfall and throughfall, as already mentioned in Sect. 2.3.1. A further factor contributing to the nutrient retention in an undisturbed ecosystem is the production of organic matter. As litter, this organic matter already has exchange capacity for ions. A certain fraction of the litter is transformed to soil organic matter, which in coarse-textured soils is the most important part of the exchange complex. It is also important in fine-textured soils, where it complexes with clay particles to form soil aggregates, increasing chemical retention by enlarging absorbing surfaces and biological retention by improving conditions for root growth. In the case of nitrogen the usually carbon-rich litter stimulates microbial immobilization of nitrogen. The wide C/N ratio in woody litter makes this process particularly important in forests.

The existence of a strong biological control of the forest nutrient cycle has been shown most convincingly by Bormann and Likens (1967, 1979) in their large-scale catchment experiments at Hubbard Brook, New Hampshire. They used herbicides repeatedly to prevent revegetation of a clearfelled area, which caused rapid leaching of both anions (nitrate) and cations from their catchments. However, more normal forestry operations also have an impact on biotic regulation (Tamm et al. 1974). Mechanical site preparations for planting, such as ploughing and rototilling, stimulate decomposition of soil organic matter and nitrogen mineralization (van Goor 1952; Lundmark 1977; Rehfuess

1981; Vitousek and Matson 1984). Even single applications of herbicides may influence nitrogen turnover significantly (Vitousek and Matson 1984). On sites with large amounts of inactive organic matter the increased mineralization may be desirable, but unless a new vegetation cover is rapidly reestablished, nutrient losses and site deterioration are likely.

Variations in intensity of biological control mechanisms are certainly important in other connections as well as after catastrophes such as clear-felling or site preparation. One early theory for biological soil formation suggested that a main difference between mull humus and mor humus was the weaker competition for nitrogen in the mull, where earthworms and other animals continuously consume fungal hyphae, important nutrient absorbers, while the fungal activity remains less disturbed in the mor layer, where the animals are much smaller (Romell 1935).

It has been generally accepted that bacteria play a dominant role as primary decomposers in mull soils, while fungal activity should dominate in mor soils. It is certainly true that fungal biomass is larger in mor than in mull (measured on a total weight or on a soil organic matter basis), but it has been shown that bacterial activity can be high also in mor (Persson et al. 1980). Most species of trees and other plants growing on both mull and mor sites have mycorrhiza, even if the mycorrhiza frequency may vary with plant species and soil horizon. The functionally important differences are hardly to be found in the proportions between bacteria and fungi. Yet there is an important functional difference between the decomposer systems in mull and mor, viz., the existence of a two-way transport of particles by digging animals in mull and the virtual absence of such transport in mor (ants do transport particles, but only on a very local scale; Troedsson and Lyford 1973). In both systems there is a transport in roots and fungal hyphae horizontally and upwards, but only of water and solutes, and a downward transport with percolating water, with a usually less important capillary rise during dry periods. Now there is an important difference between the two groups of decomposers: bacteria depend entirely on the substrate available in their close vicinity, while fungal hyphae may transport nutrients to sources of carbon (energy) and vice versa over distances of centimetres, decimetres, or even more. The roles of fungal and bacterial decomposers may therefore differ between mull and mor. It might, for instance, be speculated that without the transport function of fungal hyphae in the mor, microsites would soon be depleted of some metabolic factor needed by the decomposers. In the mull, of course, digging animals would help to renew the substrate.

A further hypothesis has been forwarded by Gadgil and Gadgil (1975), viz., that the presence of living mycorrhizal roots retards litter decomposition. This effect, if widespread, could explain the often remarkably strong visual effect of root isolation by trenching (Fig. 2.4; Toumey 1928).

While the existence of a strong biological control over nutrient leakage from natural ecosystems is commonly accepted, there is no full agreement on the roles of some of the mechanisms discussed above (which are not mutually exclusive). On the other hand, these mechanisms have not been proven to be wrong for the cases for which they were suggested. They can thus be considered as examples of possible biological control mechanisms in ecosystem regulation.

As mentioned briefly in Sect. 2.3.2, Dahl et al. (1967) have found a striking positive correlation between the (total) nitrogen concentration in the humus layer on the one hand and, on the other, its base saturation, in Norwegian coniferous forests of widely differing site quality and vegetation type (Fig. 3.2).

Similar relationships have also been found in other studies and are interpreted by Dahl (pers. comm.) as being a result of equilibria, established by nature's regulatory mechanisms. In contrast to the conditions in relatively unpolluted Norwegian forests, there are high nitrogen concentrations at low base

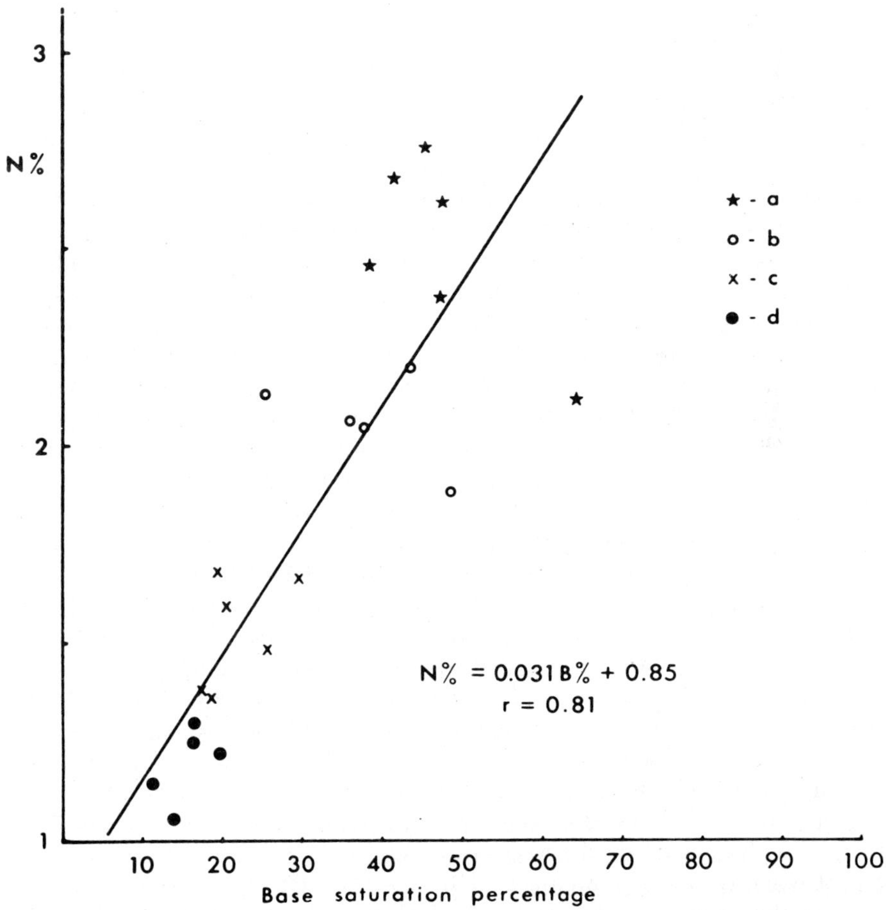

Fig. 3.2. Total nitrogen in percent of loss on ignition plotted against base saturation percent for humus samples from Romedal Almenning, Hedmark, Norway. *Symbols*: ∗ Tall herb spruce type; ○ eutrophic low-herb spruce type; x oligotrophic *Vaccinium myrtillus* spruce type; ● pine type (Dahl et al. 1967)

saturation in the Black Forest in Germany, although both vegetation types and forest stands are comparable with those studied in Norway, and it is suggested that this is due to atmospheric deposition of both nitrogen compounds and acid substances (Aune et al. 1989). The nitrogen deposition should increase the ratio nitrogen: base saturation, the sulphur deposition should increase leaching of base cations and thus decrease the base saturation.

3.4 Nitrogen Depletion by Fire and Other Disturbances

A common cause of nitrogen depletion is fire, which volatilizes most of the nitrogen in the burnt or charred organic matter. In very intensive fires the loss may amount to almost all the nitrogen in above-ground biomass, litter and top organic horizon (Ao). In less intensive fires only above-ground biomass, or part of it, together with the loose litter, is consumed. In both cases, however, nutrient retention by root uptake is interrupted for a certain period, and soil nitrification is stimulated both by accumulation of ammonium nitrogen and by the pH increase caused by the ashes (or rather by the destruction of acid organic matter).

A fire thus creates a situation where some nitrogen is lost directly, and another part of the soil nitrogen is transformed to nitrate and thus liable to leaching and denitrification. The ecological importance of these losses depends on the fire intensity, on the distribution of the nitrogen between biomass and soil in the ecosystem, on weather conditions (especially rainfall), and on the rate of vegetation recovery. The disposition of the soil to nitrification and denitrification is also important. In certain ecosystems nitrogen fixation increases after fire, and the losses may then be replaced wholly or partly. Also, changes in ground cover, e.g., to grassy vegetation and humus type (from typical mor towards moder or mull) may enhance humus storage within A_1 layers instead of in A_0 organic layers (Olson 1958).

In summary, ecosystems with a large proportion of their nitrogen store in biomass and top organic soil horizons run a great risk of losing nitrogen after fires, especially if the fires recur frequently. This applies to many tropical, subtropical, and Mediterranean woodlands and shrublands, not only in dry climates. Plant species with an efficient internal nutrient cycling, removing most of their nitrogen from leaves before the dry season and storing them in below-ground organs or in fire-resistant tree stems, have a competitive advantage also in many other fire-prone ecosystems such as temperate grasslands and tropical savannas (cf. Ellenberg 1977).

Northern coniferous forests are also adapted to fire in their own way. They usually have much of the site's store of total nitrogen down in the soil, although in a form not easily accessible to plants. Much of the cycling nitrogen is stored in the biomass of often fire-resistant conifer species (many pine species, Douglas fir, etc.) and in a mor layer on top of the mineral soil, a layer which is cold and relatively moist, at least at the beginning of summer. Under these conditions fire in the dry litter layer produces water vapour, forming a condensation zone which prevents the fire from progressing downwards, thus avoiding a total consumption of the forest floor (Uggla 1957).

On the other hand, trees and lesser vegetation with symbiotic nitrogen fixation are not very common in the boreal forest, not even after forest fire. When large losses of nitrogen occur, as in fires under very dry conditions, or when an area burns twice within a short interval, the losses are not easily compensated for. Such forests are often open, consisting of trees with some kind of fire adaption (*Pinus sylvestris* with thick bark, *Pinus banksiana*, *P. contorta*, and *Picea mariana* with serotinuos cones or other mechanisms for rapid recolonization after fire). The lesser vegetation is dominated by dwarf shrubs and in the bottom layer lichens (*Cladonia*, *Cetraria*, *Stereocaulon* and others). Both dwarf shrubs and lichens are destroyed by fire but come back, with the rate of recolonization varying strongly with species.

In Europe mature pine trees usually survive a fire, and *Calluna vulgaris* is a rapidly recolonizing dwarf shrub. Other dwarf shrubs are slower, as are also many lichens. In northern Sweden the boundary of a forest fire may still be easily distinguishable 50 years later, from the frequency of the slow-growing lichen *Cladonia alpestris*. Boreal forests in other regions have partly or entirely other species composition, e.g., in many Canadian boreal forests open stands of *Picea mariana* or *Pinus banksiana* with lichens in the bottom layer.

With an open canopy and slow-growing field and ground layer vegetation, the litter production is low and the mor layer thin. The openness and the high frequency of lichens makes this forest type inflammable, and studies in northern Sweden (Zachrisson 1977) indicate a fire frequency of one to two times per century, even in areas with a very sparse human population until 100–200 years ago. The main cause of these forest fires should then have been lightning. The frequency of lightning fires varies considerably between different parts of the circumpolar belt of northern coniferous forests. It is very high in some parts of the Rocky Mountains in North America, where thunderstorms without much rain are common.

Increased activity by man has increased forest fire frequency in the boreal zone as in many other regions, but more recently the improvements in fire combat have limited the areas affected. On the other hand, timber exploitation may lead to a nutrient depletion of the boreal forests. In the early stages of Scandinavian forest exploitation only large logs were valuable enough to offset the cost of transportation to remote sawmills. Selection cuttings of the largest and best growing trees led to open forests with similar characteristics as those mentioned for the sites with high fire frequency, including inflammability. It is often difficult to distinguish between "primary lichen-pine forests" (those due to high fire frequency over centuries or millennia) and "secondary lichen-pine forests", created by man by selection cutting, often followed by fires. Yet the primary type, which has had a thin mor layer for a long time, perhaps since the last glaciation, usually has a thin bleached horizon (A_2). The secondary type, which has replaced a moss-rich forest with a thick mor layer, usually has a bleached horizon of 5 to 10 cm depth, which is normal for moss-rich north-Scandinavian coniferous forests (see O. Tamm 1950 and references quoted there).

Fig. 3.3. Two sites situated less than 100 m apart, and with the previous stand burnt down in the same forest fire in 1933. Site **A** is on fine sand, and the humus layer contains 400 kg nitrogen ha^{-1}, a normal figure for northern Swedish *Myrtillus* type forests. On site **B** the humus layer only contains 150 kg nitrogen, meaning an impeded nutrient cycling between soil and stand. The mineral soil is a glacial till with sand and fine sand as dominating fractions, with some of the finer particles washed away from the surface soil during the postglacial land upheaval. This soil is not considered poorer than the fine sand in **A**, and a comparison between actual forest growth and that estimated from measured site properties, including texture (Hägglund and Lundmark 1977), shows that stand **A** grows somewhat better than expected from site properties, while stand **B** is much less productive than expected. No other explanation for this difference has been offered than differences in fire intensity in 1933 and possibly on earlier fire occasions. Location of site: 65° 30′N latitude, altitude 140 m, near Manjärv in the north Swedish inland (Photo C. O. Tamm September 1978, site information from J. E. Lundmark, pers. comm.)

As already suggested, a forest fire may affect site productivity and nitrogen cycling negatively or positively, depending on circumstances. Figure 3.3 A, B provides a striking example.

In late successional stages of the boreal forest (without fire) there is, at least on cold and moist sites, an immobilization of nitrogen in a thick and biologically less active mor. A forest fire or a prescribed burning then stimulates decomposition and much of the immobilized nitrogen becomes available again. Given that vegetation colonization by pioneer plants and trees is rapid enough to prevent larger losses of nutrients, nitrogen in particular, the fire has started a new succession with considerably higher primary production than that of the old growth (Sirén 1955). Modern man has imitated this process by prescribed burning, but in the last decades most prescribed burning had been replaced by soil scarification or even ploughing, at least in northern Scandinavia. As mentioned earlier, mechanical disturbances also have stimulating effects on decomposition and nitrogen release. In some forest ecosystems, fire as well as soil disturbance favours colonization by alder, legumes, or other species with symbiotic nitrogen fixation. This is seldom the case on sandy tills or sediments in northern Scandinavia, so the nitrogen lost there cannot easily be replaced in this way (Huss-Danell and Lundmark 1988).

Fire plays a less dominant role in temperate broad-leaf forests than in the forest biomes north and south of them (Olson 1981). Yet fire impact cannot be neglected, not even in the hardwood forest. The widespread occurrence of many pine species, often in relatively pure stands in the temperate zone, depends to a large extent on more or less frequent fires. In Eurosia as well as in North America fire has been a tool for humans to make land more productive with respect to food and fodder, at least since neolithic time (Iversen 1949). In Western Europe repeated burning created large areas of open heathlands, providing grazing for the survival of hardy sheep and cattle year-round. The repeated burning of these heathlands certainly meant a nutrient depletion, both of nitrogen and of cations. Some of the nitrogen could be replaced by symbiotic fixation (*Genista*, *Lotus* and other legumes), but the total store of nutrients in the root zone is usually low and, according to Rennie (1955), often not sufficient to support a well-growing forest generation.

As hardwood forests on the whole have more nitrification than northern coniferous forests, most types of human disturbances such as forest harvests, temporary agriculture and other forms of soil cultivation are likely to lead to losses of nitrogen and other nutrients, possibly larger than those in the boreal forest (but see below regarding compensating mechanisms).

In addition to the losses we have already discussed, by harvests and leaching, losses with erosion of top soil are common in many areas, in mountains as well as in less broken terrain, such as in the southeastern United States, where temporary cultivation and grazing caused large-scale erosion, both sheet erosion and gully erosion. Opening of stands by fire, grazing, or cutting on exposed sites, such as windy hilltops, may hasten downhill blowing of litter (Olson 1958), with consequences as will be discussed next.

3.5 Nitrogen Depletion by Removal of Biomass or Litter

A particularly severe form of depletion of nitrogen and other nutrients consists in the repeated removal of plant biomass or "necromass" (dead material) from the sites other than just tree stems.

The best-studied case is the practice of litter removal, which was common in Middle Europe for centuries (Rehfuess 1981). The litter was raked for use in cowsheds and stables, for mixing with the manure. Part of the humus layer and vegetation usually went with the litter. On poor sites the result became

Fig. 3.4. Litter raked and stacked for drying before removal, with the tools used for the purpose (Photo H. Holmen 1967, Bavaria)

slow-growing forests with a sparse field layer of dwarf shrubs and a ground layer of lichens and some mosses. The site quality decreased: some of the fertility of the original forest land had been transferred to the cultivated fields, a very desirable effect for the farmer in the time before chemical fertilizers.

In the areas along the North Sea, where much of the forest land had been converted to open heathland, there was less litter available, and, instead, a system was developed for moving heathland turf to arable land, creating what in German is called *Plaggenboden*, – soils with a much deeper topsoil (A_1) than normal ploughed soil. Turf removal has occurred also in other areas, mostly as a local phenomenon (e.g., collection of cover material for charcoal stacks). A method for shifting cultivation on peatlands in Finland (Finnish *kytöviljely*, Linkola 1987) involved drainage operations, but above all breaking up the surface peat by hoes, sometimes piling turfs in order to make them dry enough to burn.

In Sect. 3.4 we discussed the use of fire to improve the grazing quality of land, temporarily or more permanently. It is clear that slash-and-burn agriculture, where crops for human consumption are produced and removed, poses even stronger nutrient stresses on sites than burning for grazing, where there is no soil working and where continuous vegetation is soon reestablished. With a sparse population, as in Finland before industrialization or in tropical countries before the more recent increase in population, shifting cultivation could be a sustainable use of land resources. The people had learned, the hard way, the necessity of extended periods of forest regrowth between short periods of cultivation. The present shortening of the regrowth periods imposed by the population pressure in many tropical countries is bound to create large areas of low-productive land, depleted of nitrogen and other plant nutrients.

For less extreme conditions, such as naturally fertile soils in south Sweden, Emanuelsson (1989) has made interesting calculations of how much nitrogen

Fig. 3.5. Bavarian pine forest on outwash sand, subject to repeated litter removal. Note the sparse ground vegetation, consisting of *Calluna vulgaris* with some *Vaccinium*, and lichens, similar to that in Fig. 3.3 B (Forstamt Schwabach, Photo C. O. Tamm 1963)

land could supply annually under different forms of agricultural use, without long-term site degradation. His results are reproduced in Fig. 3.6. The figures are consistently higher for fodder-producing land (also subject to pollarding or other leaf collection) than for cultivated fields. Emanuelsson's estimates agree reasonably well with actual measurements of the nutrient cycling in a partly wooded meadow in central Sweden (Sjörs 1954). These results give a natural explanation for the dominant role of animal husbandry in the old land use, as documented in tax records and other historical evidence from different parts of Europe.

In modern mechanized forestry, harvesting often removes more of the tree from the site than the stems. As already mentioned, the stemwood has low nutrient concentrations compared with other parts of the tree, e.g., 1/20 of the nitrogen concentration in foliage, or less. Whole tree harvest means considerably higher nutrient losses to the site than stem harvest only. The shorter the rotation, the larger will be the proportion of foliage, branches, and bark in the removed above-ground biomass. There has thus arisen considerable concern about nutrient depletion with whole tree harvesting on poor sites in Scandinavia and elsewhere (Ågren 1986; Rosén 1988). Increased atmospheric deposition of nitrogen will of course compensate for some of the nitrogen losses but

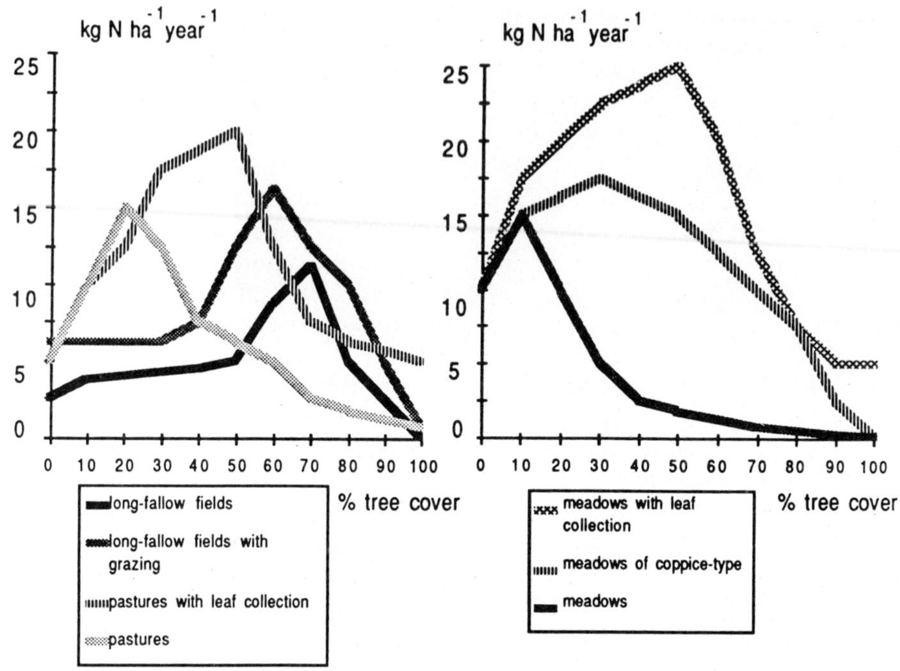

Fig. 3.6. Relationships between tree cover and annual amounts of nitrogen that over periods can be taken out of a hectare of land by harvests under different land use regimes (long-fallow fields, grazing, pollarding or coppicing, or hay-making). The diagrams are based upon available taxation records from the late 17th century, from a fertile area in southern Scandinavia, but the recalculation to kg nitrogen and the generalizations over the full scale of canopy cover necessitate certain assumptions (Emanuelsson 1989)

may lead to an unbalanced nutrition, as there is no compensation for lost cations or phosphorus.

Removal of stumps as well, with at least part of the coarse roots, has even stronger ecological implications than removal of above-ground parts. Such operations are not entirely new, formerly used in the practice of tar burning, mostly on a local scale. A total tree harvest (including stumps) is now considered a possibility of obtaining more wood for pulping and chipboard production. Such operations mean not only nutrient removal but also considerable mechanical disturbances of the soil. Organic matter decomposition and nitrogen mineralization increases. Leaching losses of nitrate as well as of other nutrients absorbed on the decomposed organic matter may then occur, in addition to the direct losses with removed biomass.

3.6 Concluding Remarks

A typical nitrogen-depleted site, as described here, is characterized by slow-growing trees, in most of Europe usually pine (*Pinus sylvestris*), with a canopy admitting a good light supply to the lesser vegetation, which, however, is not very dense and consists of species with a low nutrient demand, such as heather (*Calluna vulgaris*), lichens and some mosses. The soil is acid and the most common soil type podzol. Of course other types of nitrogen-depleted sites occur. *Calluna* heathlands have already been mentioned. Oak, birch, and beech forests may also show signs of nutrient depletion beyond that corresponding to an originally low nutrient availability. Most broad-leaf forests on medium-rich to rich soils would, however, have compensating mechanisms for at least nitrogen depletion (in biological fixation), so a site deterioration would have to pass the longer way over soil acidification → cation depletion → poor conditions for both symbiotic and free-living nitrogen fixers. Now, when widespread soil acidification occurs in central Europe, conditions for nitrogen fixation are no doubt impaired. One of nature's regulatory mechanisms is overruled by industrial operations.

If nitrogen depletion, as described above, represents somewhat extreme conditions, whether human-induced or natural, nitrogen limitation is a very common phenomenon. Primary production, and consequently the whole food web in an ecosystem, is always limited by some ecological factor, usually by more than one factor on each occasion. In addition there may be endogenous limitations in the plants themselves. It is common knowledge that the rate of photosynthesis can always be changed by altering one or more or the factors like incident light, temperature, and water availability. It is equally true that lack of one or more plant nutrients may restrict primary production, either because the element is directly involved in photosynthesis (e.g., magnesium) or because the plant has no sink for the photosynthetic products, if lack of nutrients restricts growth and metabolism.

If we try to analyze the various types of limiting factors and the feedbacks between them and the living parts of the ecosystem, we find that factors such as sunlight, temperature, and precipitation are mainly regulated externally to the ecosystem. It is true that the structure of the ecosystem affects the distribution of light, temperature, and moisture within the system, but not the total amount of sunlight received. The evapotranspiration is of course largely controlled by vegetation, but incident precipitation is only marginally affected by the vegetation (by affecting the roughness of the surface) except in special cases such as "cloud forests". The feedbacks on local, regional, and global levels by vegetational control on evapotranspiration and albedo should not be forgotten, but for most purposes we can consider the physical factors as external driving variables in the local ecosystem.

Nutritional factors are different in this respect. Of course there are also aspects here providing analogies to the external variables just discussed, e.g., the mineralogy of the parent material, which is not influenced by the biota. However, among the soil-forming factors (Jenny 1941), we find organisms, which means that there are strong feedbacks in the system, as discussed above in connection with biological retention. Vegetation and soil organisms control, to a large extent, the amounts of nutrients cycling between plants and soil (Table 2.1), and as nutrients are often limiting, evolution has favoured such organisms that are effective in extracting and/or retaining the nutrients they demand.

In evolution, high efficiency in some respect has a cost, which means that the most competitive species are usually not the best in all respects. There is often more than one solution to a specific problem, leading to the establishment of different niches in the ecosystem. One example: phosphorus is often a limiting factor in the tropics, to a large extent because soils rich in aluminium and iron (sesquioxides) fix phosphorus in a form not easily available to plants. Plants then have options, if that somewhat anthropomorphistic expression is allowed, e.g., to develop a more efficient root system, to associate themselves with other organisms more efficient in extracting phosphorus (mycorrhizal fungi), or to exploit some unconventional source of phosphorus (carnivorous plants). The various options all have their costs, and the most competitive solution is probably not the one which puts all efforts in phosphorus uptake efficiency at the expense of most other functions. Indeed, balances may be created where a mycorrhizal plant is still phosphorus-limited and dependent on its fungal partner, which in turn depends on photosynthetic products from the plant. The above-mentioned is of course not specific for phosphorus limitation, but may as well apply to nitrogen or other nutrients (Alexander 1983).

In terrestrial ecosystems, nitrogen differs from the other plant nutrients in that the amount of cycling in the ecosystem is entirely determined by its previous history in terms of acquisition of nitrogen by organisms (N_2 fixation and retention). Although there is a nitrogen pool in the bedrock (Table 1.1), most soil parent material can for all practical purposes be considered as free of nitrogen.

The strong links between ecosystem development and nitrogen store and cycling seems to be the background for the nitrogen limitation of so many types of ecosystems, arctic, boreal, temperate, and at least some Mediterranean and tropical ones. The degree to which nitrogen limits the growth is variable, and so are the regulatory mechanisms by which the nitrogen limitation is maintained. Fire in a tropical savanna may dissipate part of the nitrogen store, making the *Acacia* trees with their symbiotic fixation more competitive than they would be without repeated fires.

As mentioned earlier (Sect. 3.2) the low decomposition rate in an old boreal spruce forest takes nitrogen out of circulation and stores it in a thick humus layer. The strong nitrogen limitation in this ecosystem allows the spruce trees to attain high age and also to reproduce (with considerable difficulty, but better than other tree species). A kind of semi-steady state is created in north Scandinavian spruce forests which have escaped fire for long periods (Sirén 1955). In a climate with very cold winters and less insulating snow cover, as in the interior of Alaska, a similar vegetation succession leads to permafrost and peat formation on flat areas (Viereck 1970). In both cases the nitrogen limitation can be temporarily removed or relieved by forest fire, so it may be a semantic question to ask whether the end stage of the succession is a stable climax or just a steady state, persisting until the next forest fire. The nitrogen limitation of the primary production, coupled with the carbon limitation of the decomposer organisms (more correctly limitation by the quality of the litter, their energy and nutrient source), can be considered an important stabilizing mechanism in many ecosystems, such as many of the boreal and temperate forests and grasslands. The extent to which the phosphorus limitation and the colimitation by both nitrogen and phosphorus availability also creates stabilizing mechanisms remains more speculative, but might be worthy of further study.

Large-scale stabilizing mechanisms of the kind suggested may lead thoughts to the Gaia hypothesis (Sect. 2.2.3). Yet it should be emphasized, as was done by Lovelock himself (1979), that the possible existence of stabilizing mechanisms is not an excuse for careless handling of natural resources. Even reasonably well-stabilized systems may have points of no return.

4 Nitrogen-Enriched and Nitrogen-Saturated Ecosystems

4.1 Definitions

The term nitrogen-enriched ecosystems is used here to describe ecosystems which have received and stored nitrogen compounds in excess of what is normal during terrestrial vegetation successions. Nitrogen has thus been transferred to the site from other ecosystems by natural processes or by humans. A temporarily ample supply of mineral nitrogen released from autochthonous material by some disturbance (fire, clear-felling) is not enrichment in this sense, while a site with nitrogen released from nitrogen-rich drift walls should be classified as enriched. Accumulation of nitrogen in peat is not an enrichment in the sense that the concept is used here, as long as the peat is autochthonous and its nitrogen content comes from nitrogen fixation and preindustrial deposition on the site. However, there are cases when nitrogen-rich peats are enriched by allochthonous material transported from adjacent ecosystems, and the sometimes intensive nitrogen mineralization after draining of such peatlands may make them behave like nitrogen-saturated ecosystems (see below).

Preindustrial deposition is not considered to be a direct source of nitrogen enrichment. The problem is that the preindustrial levels of deposition are unknown. As mentioned in the introduction, studies of ice cores from southern Greenland seem to indicate a twofold increase in nitrogen deposition as well as in that of sulphate in this remote area from 1885 to 1978 (Neftel et al. 1985). As the precursors of deposited nitrate have shorter residence times in the atmosphere than those of sulphate, it seems to be a safe conclusion that deposition has increased with a higher factor near industrial source areas, and that preindustrial nitrate deposition was only a fraction of the present one over large parts of Europe (cf. Skeffington and Wilson 1988). We do not have corresponding data for ammonium deposition, except for a few old series of rainwater analyses of variable quality. The measurements from Rothamsted mentioned in the introduction probably constitute one of the more reliable series, and do not show a very clear trend for ammonia wet deposition (Brimblecombe and Stedman 1982). As much of the ammonia emissions originate from animal husbandry (manure, urine), there was probably considerable emission also in the past. What seems clear is that the intensity in modern animal husbandry, with its large numbers of animals, and with much protein-rich fodder used, has increased the ammonia deposition regionally.

According to one definition, nitrogen saturation is considered to occur when primary production is not further increased by increased nitrogen supply (Nilsson 1986). This definition relates to physiological concepts (some kind of optimum curve for the dominant primary producers in the ecosystem), and includes a number of cases where some other factor (water, or nutrients other than nitrogen) limits primary production so severely that even low amounts of nitrogen are sufficient. Therefore, the term nitrogen saturation is also used in a more restricted meaning, i.e., ecosystems where nitrogen losses exceed nitrogen inputs, measured over long periods (Ågren and Bosatta 1988). This latter definition also refers to physiological processes, in particular to the retention of nitrogen by biological immobilization. The retention is certainly higher in ecosystems far from nitrogen saturation in the first sense than in those at or above optimum, but there is no full congruence between the two definitions. Nitrogen retention has, for instance, a temporal variation, both with season and with successional stage of the vegetation (Vitousek 1982). A definition implying that outflow of (nitrate) nitrogen should equal or exceed atmospheric input could be misleading, if much of the nitrogen losses are by denitrification. Awaiting scientific agreement on definitions, we shall here follow the recommendation by Skeffington and Wilson (1988), that each author using the concept has to give a definition: we consider nitrogen-saturated such ecosystems where the physiological nitrogen demand of the primary producers is satisfied, and where considerable nitrate leaching occurs.

It may still be a matter of discussion regarding how to label ecosystems with a temporarily high availability of nitrogen, leading to leaching losses. If it is a question of seasonal variability, such as winter leakage, we can resort to the annual balance input/output. In the case of successional stages, such as high nitrogen availability on cleared forest areas, leading to high leaching losses, some but not all forest ecosystems may fit some definitions for nitrogen saturation for a period of one to several years, even if they are unsaturated over the rest of the rotation period.

It should be noted that nitrogen-enriched ecosystems are not necessarily nitrogen-saturated. Nor is it always true that nitrate leaching must be a sign of nitrogen saturation, viz., if the leaching is temporary and a consequence of disturbance on sites with high stores of organic nitrogen compounds built up over a long time, by enrichment or nitrogen fixation.

4.2 Sites Naturally Enriched with Nitrogen

Natural causes for nitrogen enrichment are, as was said in the previous section, transfers from other ecosystems, adjacent or farther away. Examples are lake and seashores with accumulation of driftwalls, alluvial sites such as seasonally flooded valley bottoms receiving both dissolved and suspended material from other parts of the catchment, flushed sites such as lower parts of long slopes receiving surface or subsurface water trickling down, and bird colonies such as those producing guano. The donor ecosystems do not need to be particularly rich in nitrogen, but there must be an efficient collecting mechanism. The foraging of sea birds over large areas in the guano example and the ability of plant roots to absorb nitrate ions and other nutrients from a very dilute (but more or less continuously renewed) solution in the soil water of a flushed site represent two very different enrichment processes, and their implications for the function of the landscape complexes differ qualitatively and quantitatively.

It can be concluded from the above that sites with strong nitrogen enrichment seldom cover large areas. The enriched sites are ecotones or otherwise special sites within the landscape mosaic on which they depend. On the other hand, nitrogen enrichment occurs also on a meso- and microscale in most terrestrial ecosystems (anthills, dung heaps, carcasses of animals of different sizes, etc.) and helps to create the soil heterogeneity which is characteristic and functionally important for natural and seminatural terrestrial ecosystems. A moderate enrichment, such as that on flushed sites, at lower ends of hillslopes (see Sect. 2.3) may affect larger segments of the terrain, but never the whole landscape.

4.3 Nitrogen Enrichment by Preindustrial Agricultural and Pastural Land Use

Human nitrogen enrichment may occur in different ways. We have earlier mentioned the nitrogen depletion caused by litter or turf removal, but the fields which eventually received the material, together with manure, were of course temporarily enriched, until harvesting had removed the extra nutrients. However, long before the practice of litter removal, herds of domesticated animals grazed over wide areas but preferentially left their dung and urine in places where they were watered, milked, or collected overnight. And humans collected food from large areas and left the residues close to settlements, where archaeologists can reveal their existence a thousand years later by analysis of soil phosphate. Nitrogen was enriched together with phosphorus, but is much more biogeochemically mobile, so differences in nitrogen concentrations are evened out more rapidly than in the case of phosphorus.

Until chemical fertilizers became available in the latter part of the nineteenth century (for nitrogen in particular from the early twentieth century onwards, when molecular nitrogen could be fixed industrially), the nitrogen enrichment due to agricultural activities followed a pattern not very different from what was said above about natural enrichment, although extending over a larger part of the landscape with increasing population density.

Quantitative estimates of the amounts of nutrients moved between landscape segments in preindustrial time are rare, but one example was mentioned in Sect. 3.5 and is illustrated in Fig. 3.6.

There was a balance needed between the areas supplying the nitrogen and that deliberately enriched, the arable land (Emanuelsson 1989). Small areas, e.g., around stables, might have extreme enrichment, leading to intensive nitrification. It was soon discovered that saltpetre, indispensable for gunpowder and always in short supply, could be prepared from leachates of the soil around and below stables, after mixing with potash. Legislation was introduced in some countries to ensure access to this valuable resource.

When the balance between land supplying grazing and winter-fodder for cattle and the arable land supplying human food became unfavorable, the agricultural system was threatened. This was the case in large parts of Europe during the nineteenth century, when population growth increased rapidly. The details of the process differed between countries and regions, but a general trend was a large increase in arable land, at the expense of land producing fodder and wood. The earlier transfer of nutrients to the fields from forest and pastures was interrupted or reduced. The situation has many parallels with what is happening in developing countries in many parts of the world at present.

The relatively balanced situation in old agriculture of course had exceptions. Some researchers may question whether there really was a balance at all, but there are so many records of a continuous agricultural use of the European landscape over centuries and even millennia that we can at least conclude that changes in nutrient balance were slow and landscape productivity could be maintained over long periods in many areas. In other cases we have evidence for rapid changes in population density and land use intensity, which often can be related to climate fluctuations, pests on man, cattle, and crops, or to human migrations. The last phenomenon might well be triggered by one or more of the other causes just mentioned, or by a misuse of natural resources, as has been suggested by historians in several cases. It is not easy to reveal the part in history played by misuse of nitrogen and other nutrients for, e.g., the migration of Europeans to other continents in the last centuries, but the overuse of natural resources is certainly one component in the complex causal relationship.

The creation and growth of large towns formed a striking exception from the relatively balanced situation in the countryside, where most people lived before the industrial revolution. Towns were large importers of both human and animal food and thus enriched to a degree where the environment became very unhealthy, to use a mild word. The high mortality in European urban communities once towns started to grow in the late Middle Ages had several causes, but enrichment with both organic material and nutrients certainly favoured pests and their vectors. The quality of drinking water deteriorated when it became contaminated with high concentrations of organic substances as well as of nitrate, another parallel to conditions now often met in towns and villages in the developing world.

4.4 Enrichment by Chemical Fertilizers and Urban Waste Products

The situation changed dramatically once agriculture had access to relatively cheap nitrogen fertilizers. Harvests increased dramatically, and so did the amount of nitrogen cycling within the farm. We know now that not only the input with fertilizers (with imported fodder and sewage sludge providing additional inputs) and the losses with agricultural products increased, but also the losses with drainage water. The possibilities of regulating nutrient levels and nutrient balances in agricultural systems have had further far-reaching consequences. The increased and more regular harvests (also a result of plant breeding) has changed the economic structure of the entire agricultural sector, starting in some industrialized countries in Europe and North America but gradually extending to many countries on all continents. This in turn has had profound ecological implications, with a development towards large specialized farms with high input of energy and material from outside, with intensive use of pesticides and little room for organisms other than those directly involved in economic production. On the other hand, in some parts of the world large areas of low-quality arable land (less commonly, fertile land as well) has been taken out of agricultural production and often converted to forest.

Although there is in many countries a critical discussion of this large-scale agriculture with its philosophy strongly influenced by industrial economics, the knowledge of the long-term effects on soils and ecosystems of modern management methods is far from satisfactory. There is a number of long-term experiments with different management regimes running, but they do not always cover the full range of treatments realized in modern agricultural practice. It has been possible in some experiments to maintain high yields in monocultures over long periods with supply of nitrogen and other nutrients exclusively in commercial fertilizers (see e.g., Johnston and Mattingly 1976). Yet it is common knowledge that soil organic matter (and total soil nitrogen) tends to decrease under management regimes with frequent ploughing. The decrease in soil organic matter is clearly undesirable, but the actual differences in yield and environmental impacts between regimes with farmyard manure and those with all nutrient replacement in chemical fertilizers depend on local conditions (Mattingly et al. 1975; Bjarnason 1987).

In grasslands managed for grazing or hay-making there also exists a certain number of long-term experiments with different combinations of fertilizers. The oldest and most famous ones are the Park Grass Plots at Rothamsted (Thurston et al. 1976), where treatments began in 1856, and where continuous fertilization with ammonium sulphate decreased the number of herb and grass species from about 60 to only a few, at the same time as the soil became acidified. However, the same amount of nitrogen as sodium nitrate decreased the number of species less, even if some of the tall grasses were favored at the expense of the low herbs common on the unfertilized plots, whether limed or unlimed. Swedish experiments (Fogelfors and Steen 1982) show that the number of species on pasture plots decreased drastically during a 25-year period on heavily fertilized plots. Steen (1980) has summarized the results from several experiments in a model showing that field layer vegetation in widely different forest types converges to a species-poor (but productive) *Poa pratensis* meadow when the land is cleared and grazing and fertilization introduced (Fig. 4.1). Grime and collaborators have also found a decrease in species diversity with increasing biomass (Fig. 4.2).

Tilman (1988) has developed a model for plant competition, based on three simple assumptions: plants compete for light supply, soil resources and are

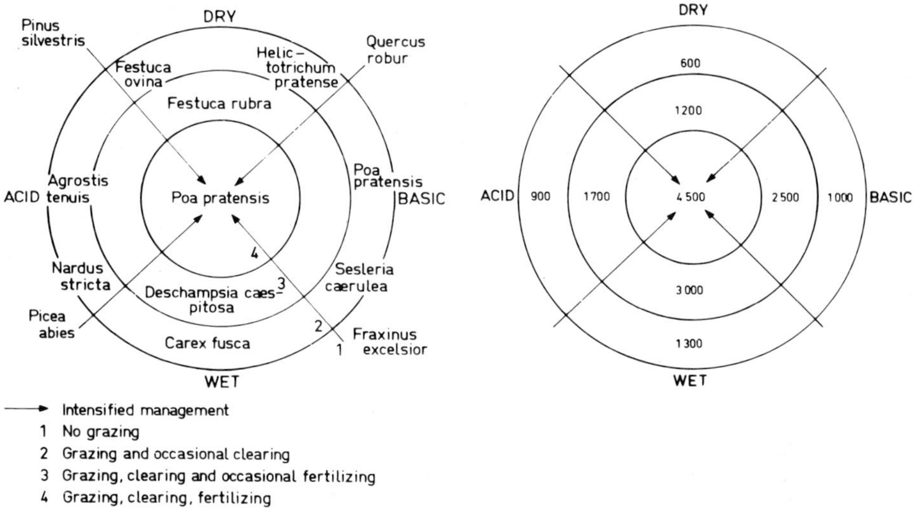

Fig. 4.1. Changes in dominating grass species and net primary production in the *field layer* when different types of Scandinavian forests are exposed to increasing human influence (grazing, clearing, and fertilization). Net primary production in *right* figure given as kg dry matter ha^{-1} $year^{-1}$ (Steen 1980)

subject to losses (by grazing and other disturbances). The success of a particular species at sites with different combinations of light, nutrient supply and loss rate is then largely determined by its allocation of resources to (1) stem growth ensuring good light conditions, (2) to roots exploiting soil resources and (3) to rapid replacement of leaves lost by disturbances. The model could elegantly simulate the development observed in large experiments in Minnesota with ammonium nitrate applications to abandoned fields in different successional stages, including uncultivated oak savannah, the presumed condition before cultivation (Tilman 1986, 1987, 1988). The experimental results fitted well with a comparative study of species competition in relation to soil nitrogen availability on old fields in the same region (Tilman 1986). In both cases species diversity decreased with increasing nitrogen level.

It has been experimentally confirmed that the present increase in dominance of the two grasses *Molinia coerulea* and *Agrostis canina* on Dutch heathlands is related to high ammonium deposition, not primarily to acidity changes (Roelofs et al. 1987). A survey of threatened plant species in Central Europe (H. Ellenberg Jr. 1985) has shown that 75% – 80% of them are characteristic for plant communities with low nitrogen availability, and that about half the flora of Central Europe is competitive only on sites with low nitrogen supply (Fig. 4.3).

Over the last few decades nitrogen fertilization has been extended from arable land and pastures to forest land in different parts of the world. Even where nitrogen fertilizers are not used, a stimulation of symbiotic fixation may occur,

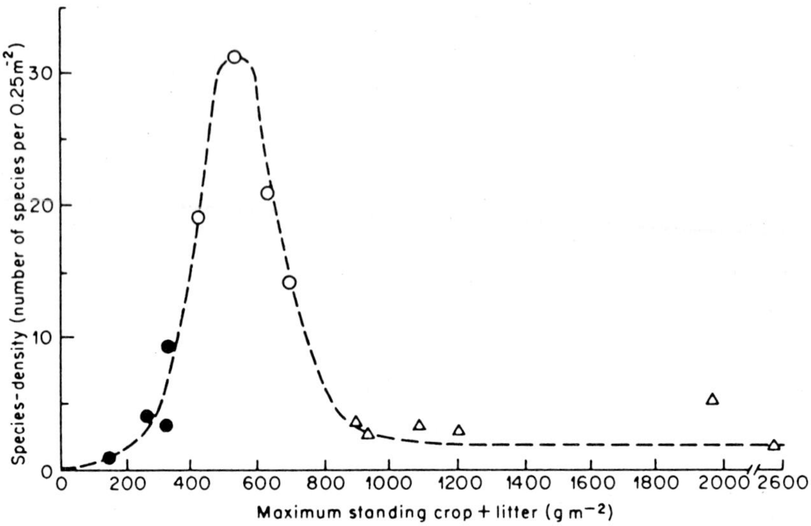

Fig. 4.2. The relationship between maximum standing crop in the field layer plus litter and species density of herbs at 14 sites in northern England. *Open dots* Grasslands. *Filled dots* Woodlands. *Triangles* Tall, herb-dominated sites (Al-Mufti et al. 1977, Grime 1979)

most often in the reforestation phase, with introduction of lupines or other legumes, or alders. Liming and phosphate fertilization are often used to stimulate these plants. The justification of the melioration measures lies of course in the result from experiments made in the past, showing growth increases after nitrogen addition on many sites, making it probable that forest management should profit from fertilization in the same way as agriculture has done.

Undesired ecological side effects of increased nitrogen supply to forests have first been observed as nitrate leaching after application of nitrates and to a lesser extent other nitrogenous fertilizers (Tamm et al. 1974). Vegetational changes are common after nitrogen application to boreal forests (Fig. 4.4, see also Fig. 2.5), which typically are rather open (Gerhardt and Kellner 1986; Dirkse and van Dobben 1989). Lichens, especially those with nitrogen-fixing blue-green algae, are disfavoured and grasses increase relatively to dwarf shrubs. In dense stands on good sites in the boreal zone and in planted conifer stands in the temperate zone the increasing density of the canopy after fertilization usually prevents development of much field layer vegetation. There are also reports that unbalanced nitrogen fertilization may lead to growth disturbances associated with deficiencies in other elements, such as boron (Raitio and Rantala 1977; Braekke 1977; Möller 1983; Aronsson 1983) and to decreased resistance to climatic or other stresses (Aronsson 1980). It is commonly known among silviculturists working in cold climates that forest nurseries should not be fertilized with nitrogen late in the vegetation period, because of the risk of winter damages, due to frost or desiccation, or a combination of both. However, it is not clear whether the nitrogen acts by retarding the hard-

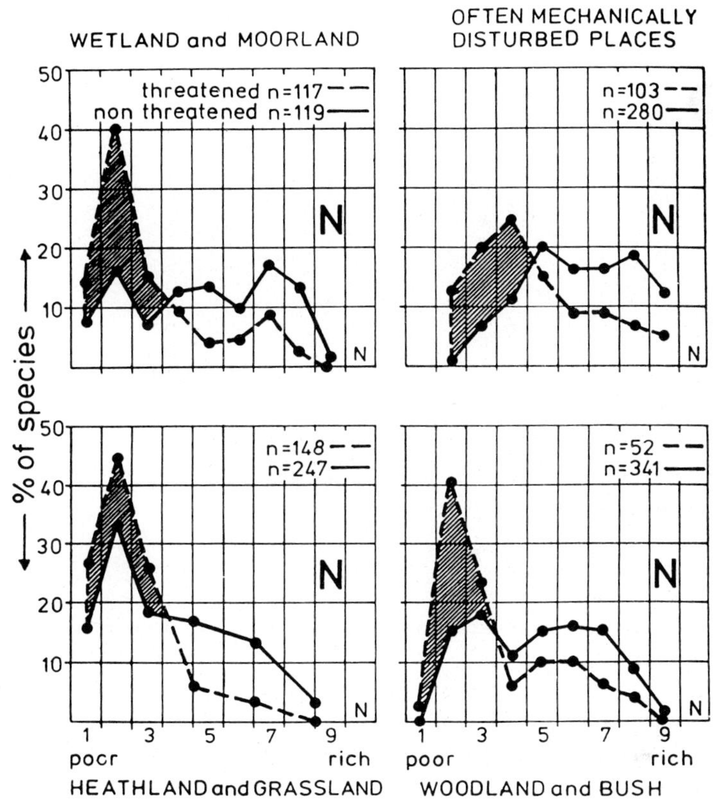

Fig. 4.3. Distribution of threatened and non-threatened Middle European plant species along a gradient 1 to 9 of increasing preference for high nitrogen availability. Most threatened species concentrate in the poor part of the gradient, while the non-threatened as a group appear more indifferent (H. Ellenberg Jr. 1985)

ening process or by lowering the concentration of soluble carbohydrates. Both soluble carbohydrates and hardening (which, i.a., includes membrane changes) are important for winter survival (Sakai and Larcher 1987). Certain amino acids, proline and arginine, are also reported to have a frost-protecting influence (Sakai and Larcher 1987).

Much of the information on nitrogen effects on forest ecosystems is derived from experiments where the applications of nitrogenous fertilizers have been made once or with intervals of several years, in which case much of the effects may be transient. There is then an urgent need for information on long-term effects on forest ecosystems of increased nitrogen input. There are of course no equivalents in forest science to the more than 100-year-old Rothamsted experiments with field crops and grassland. However, there are a few experiments done where fertilizer additions have been repeated frequently over some years (Miller et al. 1979; Weetman and Fournier 1984). The Swedish optimum nutri-

Fig. 4.4. Reaction of trees and other vegetation to irrigation and fertilization (*left* of trail) on a nutrient-limited site in middle Sweden. Note denser pine crowns and dominance of *Epilobium angustifolium* on treated area, while the field layer on untreated plot (*right*) consists of scattered *Calluna vulgaris* and *Vaccinium vitis idea* and the bottom layer is dominated by lichens. Photo taken in August 1977, after 3 years of treatment. For further details, see Persson (1981)

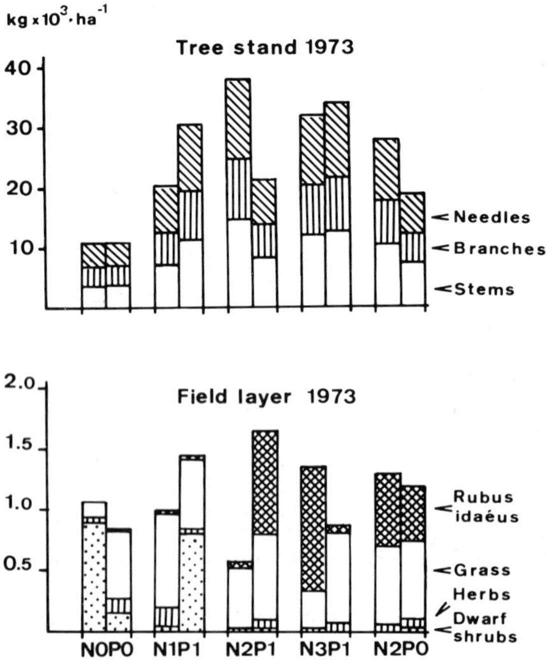

Fig. 4.5. Above-ground biomass of spruce stand and vegetation in five of the treatments of Experiment E 26A Stråsan in 1973. Note the replacement of dwarf shrubs with the nitrophilous *Rubus idaeus* at higher nitrogen levels (Albrektson et al. 1977)

Table 4.1 Fertilizer additions (kg elements ha^{-1}) in experiments E 26A Stråsan and E 55 Norrliden

E 26A[a]	N_1	N_2	N_3	P_1	P_2	K	Mg	Ca	B	Co	Zn	Cu	Mn	Mo
1967	60	120	180	20	40	80	22	4000	0.23	0.01	0.20	0.34	0.77	0.01
1968	60	120	180	–	–	–	–	–	–	–	–	–	–	–
1969	60	120	180	20	40	80	22	–	0.80	0.05	0.72	1.20	2.70	0.05
1970	40	80	120	10	20	–	–	–	–	–	–	–	–	–
1971	40	80	120	–	–	–	–	–	–	–	–	–	–	–
1972	40	80	120	–	–	–	–	–	–	–	–	–	–	–
1973	40	80	120	–	–	–	–	–	–	–	–	–	–	–
1974	40	80	120	20	40	80	22	–	0.58	0.04	0.51	0.86	1.93	0.04
1975	40	80	120	–	–	–	–	–	–	–	–	–	–	–
1976	40	80	120	–	–	–	–	–	–	–	–	–	–	–
1977	30	60	90	20	40	80	22	–	1.17	0.07	1.03	1.72	3.86	0.07
1978	30	60	90	–	–	–	–	–	–	–	–	–	–	–
1979	30	60	90	–	–	–	–	–	–	–	–	–	–	–
1980	30	60	90	20	40	80	22	–	1.17	0.07	1.03	1.72	3.86	0.07
1981	30	60	90	–	–	–	–	–	–	–	–	–	–	–
1982	30	60	90	–	–	–	–	–	–	–	–	–	–	–
1983	30	60	90	–	–	–	–	–	–	–	–	–	–	–
1984	30	60	90	20	40	80	22	–	2.50	–	–	–	–	–
1985	30	60	90	–	–	–	–	–	–	–	–	–	–	–
1986	30	60	90	–	–	–	–	–	–	–	–	–	–	–
1987	30	60	90	–	–	–	–	–	–	–	–	–	–	–

E 55[b]	Ammonium nitrate			P_2	K_2	B	Urea		
	N_1	N_2	N_3				N_1	N_2	N_3
1971	60	120	180	40	75		60	120	180
1972	60	120	180				60	120	180
1973	60	120	180				60	120	180
1974	40	80	120	40	75		40	80	120
1975	40	80	120				40	80	120
1976	40	80	120				40	80	120
1977	30	60	90	40	78		30	60	90
1978	30	60	90				30	60	90
1979	30	60	90				30	60	90
1980	30	60	90	40	78	2.5	30	60	90
1981	30	60	90				30	60	90
1982	30	60	90				30	60	90
1983	30	60	90	40	78		30	60	90
1984	30	60	90				30	60	90
1985	30	60	90				30	60	90
1986	30	60	90	40	78		30	60	90
1987	30	60	90				30	60	90

[a] At Stråsan all nitrogen was given as ammonium nitrate and phosphorus as superphosphate. The elements K and Mg were given together from 1967 – 1980 after addition of a mixture of the six last elements. In 1986, fertilization of the treatments given N_1P_1 and N_3P_1 (with and without KMg, etc.) was discontinued.

[b] At Norrliden nitrogen was given as urea (U) or ammonium nitrate (AN) and P and K were combined, given with a commercial PK fertilizer.

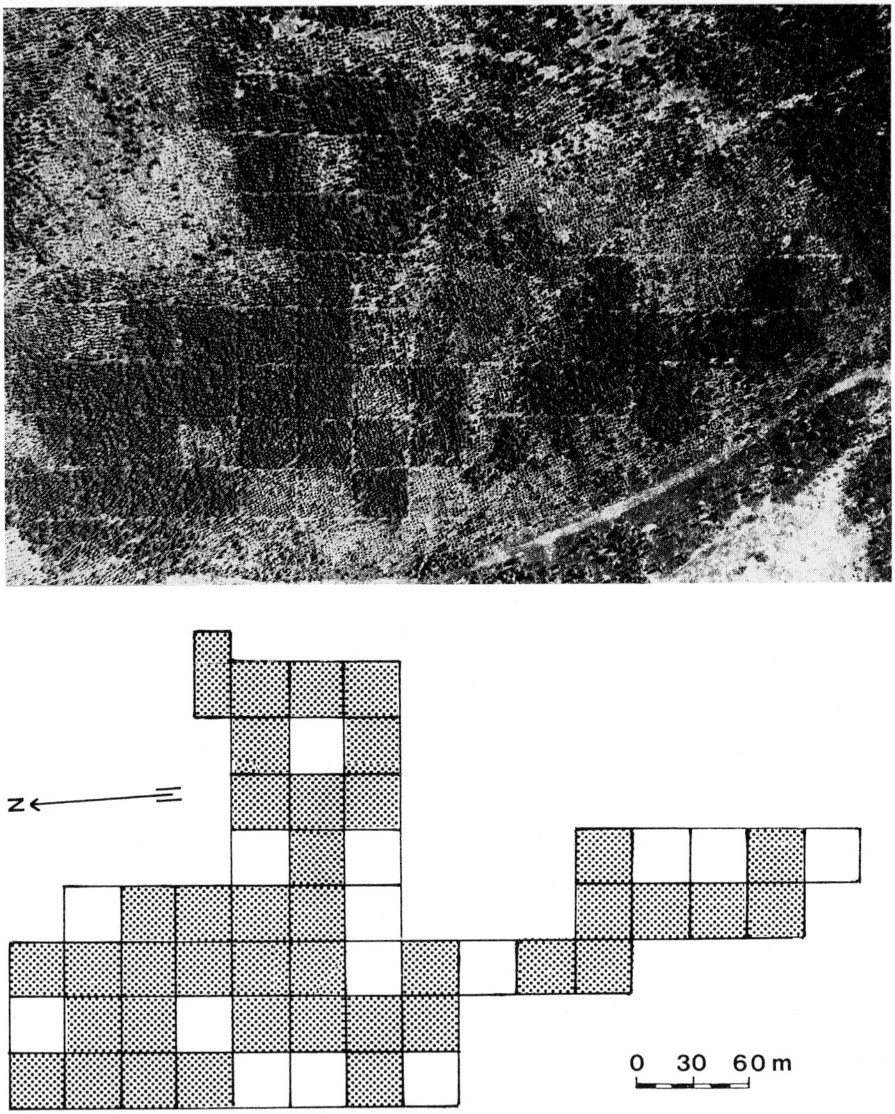

Fig. 4.6. Aerial view of Experiment E 26 A Stråsan, July 1975, showing that all plots fertilized with nitrogen (see plan at bottom of picture, where plots with nitrogen are *shaded*) show up as *dark quadrates*. Photo published with permission from the Swedish national defense

tion experiments in young pine and spruce stands have now run for periods of 15 to 20 years, with different nitrogen regimes, where at least the lower ones are comparable to present nitrogen deposition in industrialized regions (Table 4.1, see also Tamm 1985, 1989). Some experiments in older stands, started in 1959 and 1963 with less intensive but frequently repeated fertilization, can pro-

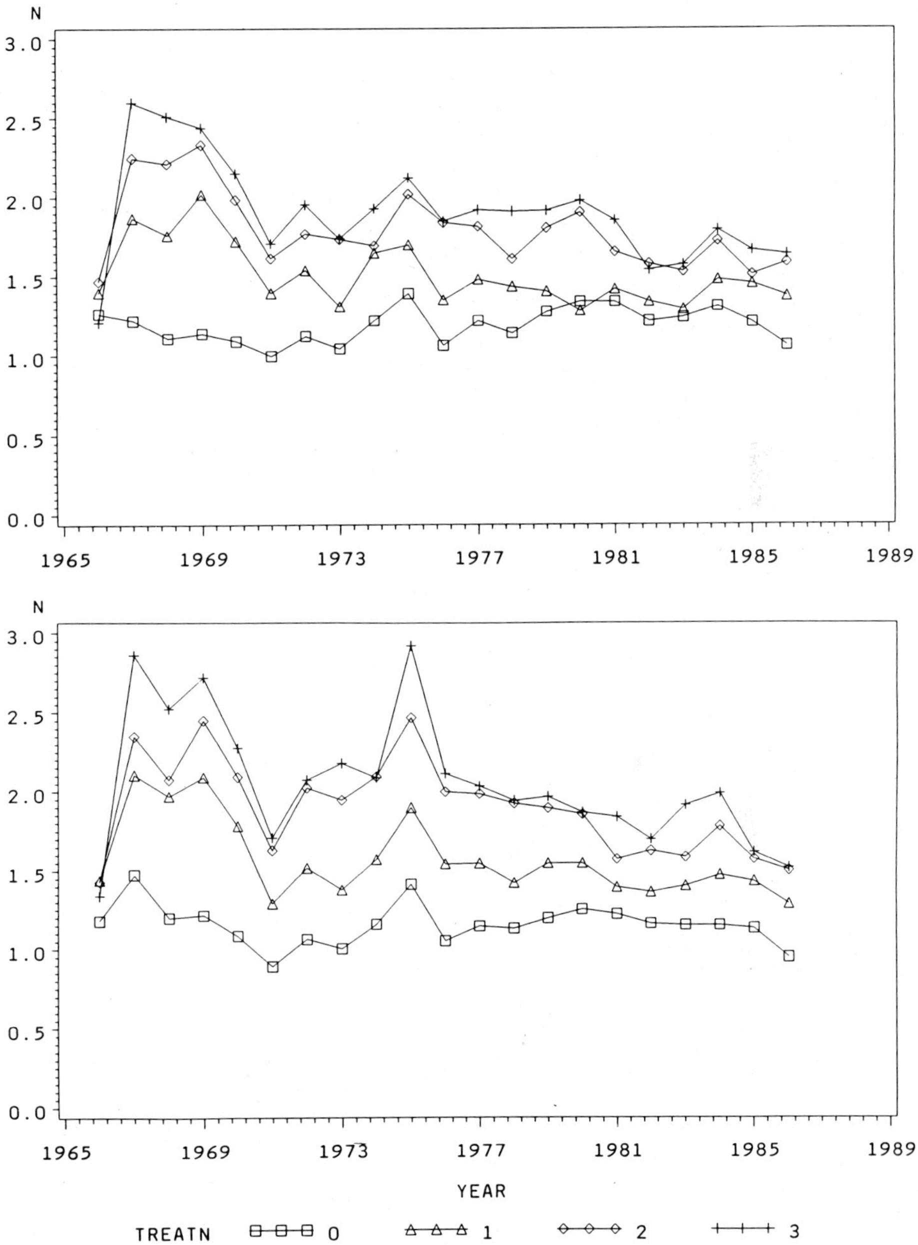

Fig. 4.7. Nitrogen concentrations in exposed current spruce needles (percent dry weight), sampled in autumn from some of the treatments in Experiment E 26A Stråsan. *Above* Only ammonium nitrate fertilization; *below* ammonium nitrate plus P2 and KMgMicro added (see Table 4.1). *1*, *2*, and *3* denote the nitrogen regimes (Tamm et al., in prep.)

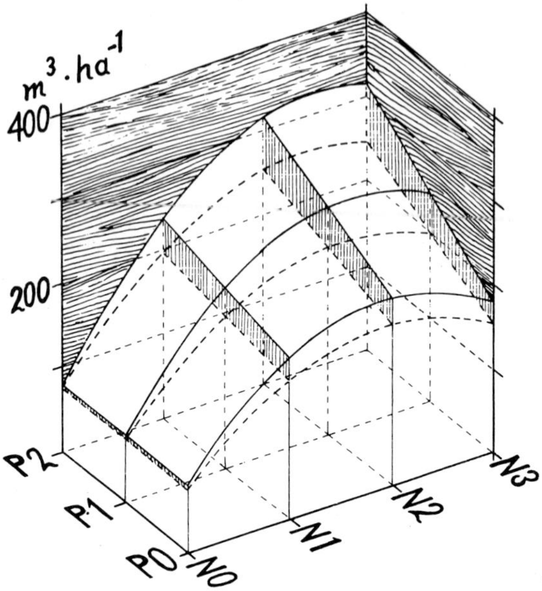

Fig. 4.8. Total stem volume production over bark in Experiment E 26 A Stråsan, as affected by fertilizer regime. The *upper surface* represents plots with KMgMicro given in addition to the N and P levels noted (cf. Table 4.1). The *dotted surface* stands for plots without KMgMicro. All treatments are duplicated once. The equations for the surfaces have the form:
$V = a_1 + a_2 \cdot B + a_3 \cdot N - a_4 \cdot N^2 + a_5 \cdot (N \cdot P)$. V denotes stem volume and B denotes the "Björgung index", the square root of the number of trees per plot times the square of their arithmetic mean height (at the start of the experiment). N and P stand for the nitrogen and phosphorus regimes as described in Table 4.1. All variables are statistically significant

vide additional information (Burgtorf 1981). We are thus not entirely lacking in information concerning the reaction of at least boreal forest ecosystems upon prolonged exposure to elevated levels of nitrogen.

The oldest plots in the Swedish optimum nutrition experiments in forests were laid out in 1967 in a young stand of Norway spruce (*Picea abies*), planted in 1957 but slow-growing and with symptoms of deficiency in nitrogen. Addition of fertilizers in a factorial design (Table 4.1) showed a rapid growth response for nitrogen (ammonium nitrate), well visible on aerial photographs (Fig. 4.6). Within a few years a positive response to phosphorus appeared, either as a main effect or as an NP interaction, depending on the growth measure and the period selected. Later, it also appeared that those plots grew better that had received a mixed application of the elements K, Mg, B, Co, Zn, Cu, Mn, and Mo, although the difference in stem volume did not attain full statistical significance in 1982 (Mead and Tamm 1988). The nitrogen levels in the spruce needles increased in the first summer almost proportional to the ammonium nitrate addition and have remained elevated on fertilized plots compared with controls (Fig. 4.7), despite a lowering of the nitrogen additions to regime N_1 equal to 30 kg N h^{-1} $year^{-1}$ (from 1977 onwards; in the

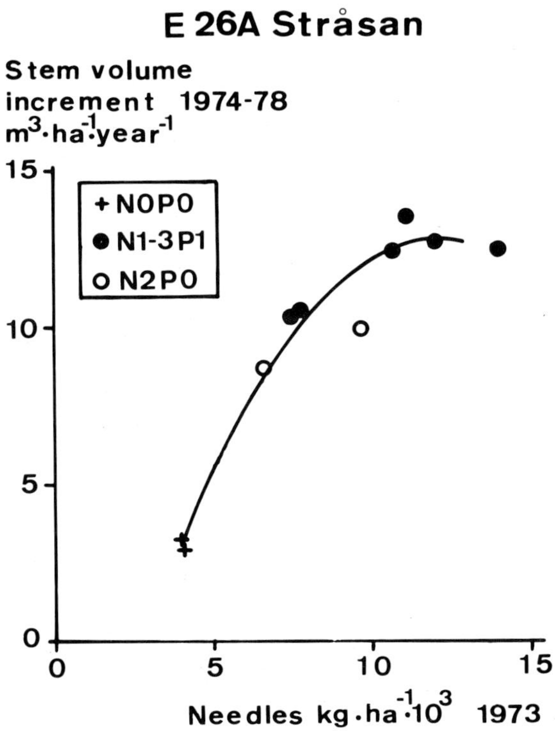

Fig. 4.9. Stem volume increment 1974–1978 in biomass-sampled plots of Experiment E 26A as a function of needle biomass at the start of the period (1973) (Tamm 1979)

first three years N_1 equalled 60 kg ha^{-1} year^{-1}). N_2 and N_3 are multiples of N_1.

The integrated growth responses (stem volume in 1986 as function of nutrient regimes) is illustrated in Fig. 4.8. The main mechanism for spruce growth response appears to have been increased allocation of photosynthate to needle production (cf. Sect. 2.1), which has led to increased growth of all other organs, at least above ground (Fig. 4.9). Above-ground production was proportional to needle biomass (Albrektson et al. 1977) until full canopy closure, which on well-fertilized plots was attained between 1973 and 1978. There was no further increase in needle biomass and consequently not in total production at the next biomass investigation (in 1982, Axelsson 1985), which accounts for the levelling off of the curve in Fig. 4.9. Although stem volume growth has increased somewhat for each measuring period, this is probably more a consequence of changes in growth allocation with tree height and lifting of the crown (Fig. 4.10) than of further increases in primary production.

The stem volume growth during the last revision periods has approximated 25 m^3 ha^{-1} year^{-1} as a mean for six plots with a well-balanced fertilization (N_2P_2KMgMicro), while both controls and other plots without nitrogen produced around 6 m^3. Plots fertilized with N_3 gave lower than maximum yield, and so did plots with less balanced nutrition, as shown in Fig. 4.8. Some plots

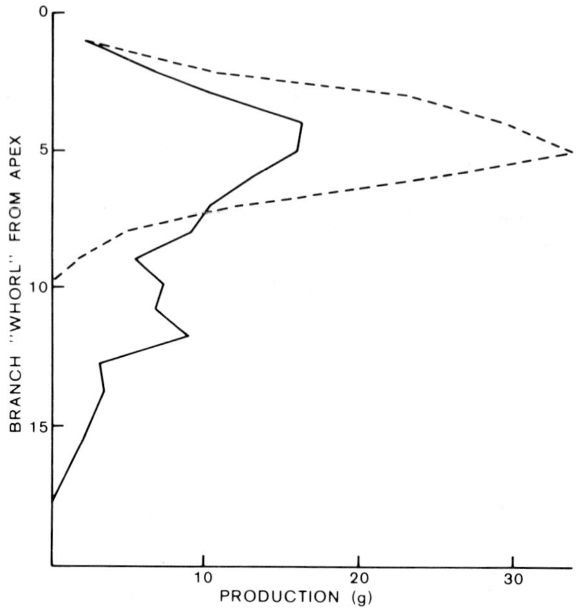

Fig. 4.10. Needle production (weight of current foliage) per branch in 1982 as related to position in crown. *Solid line* unfertilized; *broken line* fertilized N_2P_2K. Experiment E 26A. The very dense canopy on fertilized plots does not permit the formation of new needles on lower branches, and also completely shades most ground vegetation (Madgwick and Tamm 1987)

with high N but low P showed deficiency symptoms, i.a., premature needle abscission. Forking and other top damages were more common on plots without KMgMicro. These damages were probably a symptom of deficiency in boron (Aronsson 1983, 1985b), as foliage boron concentrations were very low.

Apparently the experimental treatments in the Stråsan experiment have created conditions which can be considered as steady-state, at least with respect to annual nitrogen supply, tree foliage levels, and tree growth over the last decade. The ranges of nitrogen regimes extend from those of the strictly nitrogen-limited boreal forest in a clean environment (wet deposition <5 kg N ha^{-1} year^{-1}) over additions of 30 kg N ha^{-1} year^{-1} (close to conditions in southernmost Sweden and large areas of Middle Europe) to twice and thrice that level (conditions occurring in the Netherlands and other high deposition areas). It is now a very relevant question to ask what the nitrogen losses are from plots at these various regimes. Studies have just begun, with the lysimeter tech-

Fig. 4.11. Nitrogen concentrations in exposed current pine needles sampled in autumn from Experiment E55 Norrliden. The *symbols* stand for different nitrogen regimes (see Table 4.1). The *broken lines* (OO) represent the means of all six control plots, while NO in *left* diagrams represents the mean of the three control plots in blocks fertilized with either urea or ammonium nitrate. In the diagrams to the *right*, NO stands for means of two PK plots (Tamm 1989)

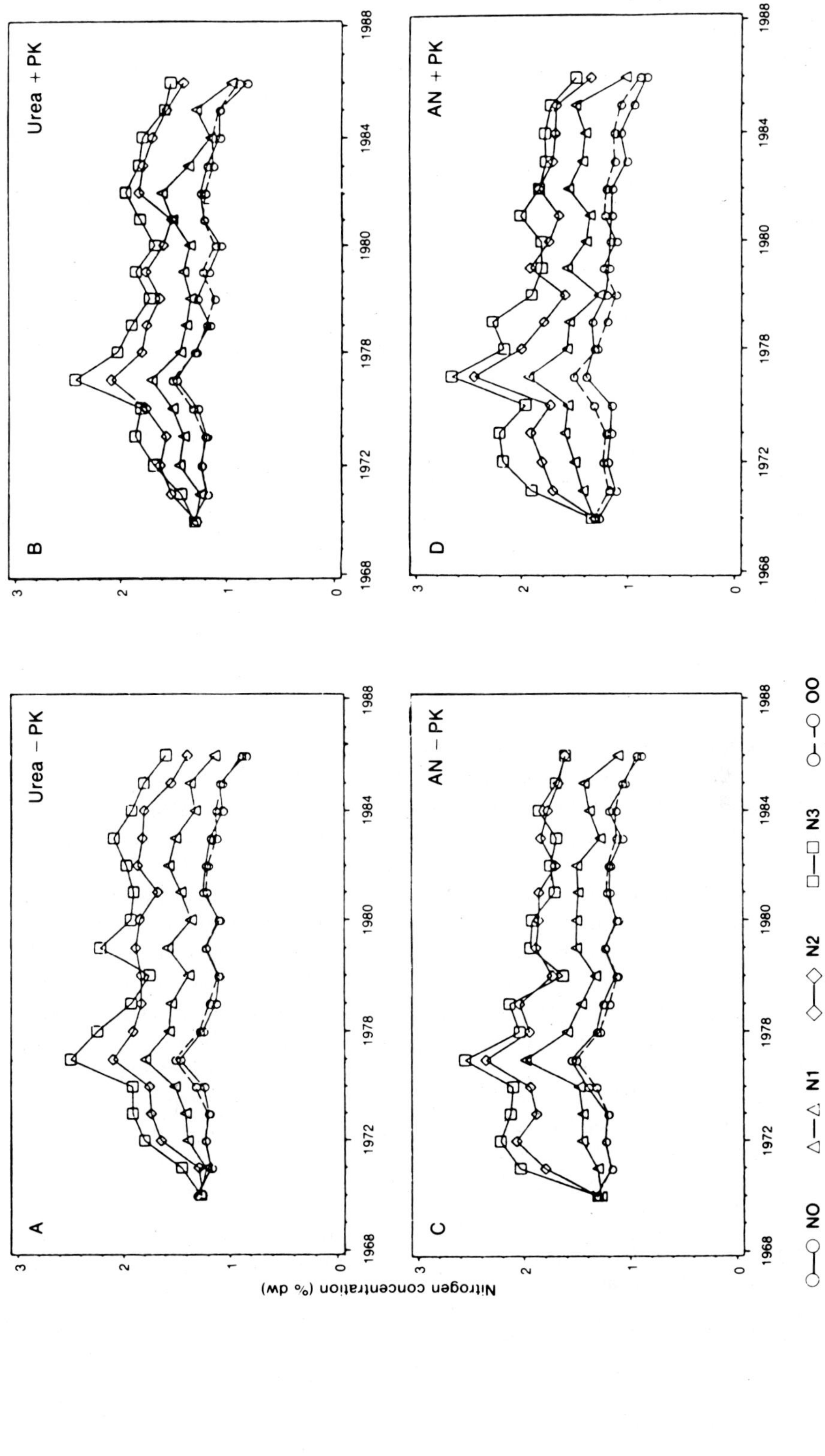
A
Urea – PK
B
Urea + PK
C
AN – PK
D
AN + PK
Nitrogen concentration (% dw)
0
1
2
3
1968
1972
1976
1980
1984
1988
NO
N1
N2
N3
OO

Table 4.2. Average nitrate reductase activity (NRA, μmol NO_2 g^{-1} h^{-1}) in *Deschampsia flexuosa* on plots in the Norrliden experiment on July 10, 1985. For treatments, see Table 4.1[a]

Treatment	NRA
Control	
(O)	0.18 (0.15 – 0.22)
Urea	
N_1	0.20 (0.18 – 0.21)
N_2	0.55 (0.49 – 0.59)
N_3	0.61 (0.55 – 0.65)
Control	
(O)	0.21 (0.15 – 0.28)
Ammonium nitrate	
N_1	0.51 (0.44 – 0.56)
N_2	0.62 (0.47 – 0.82)
N_3	0.63 (0.58 – 0.70)
Lime	0.23 (0.20 – 0.26)

[a] Values in parentheses give the range of activity ($n = 3$); from Högberg et al. (1986).

nique but also with traditional incubation studies. So far the results indicate that nitrification has not been strongly stimulated by the treatment N_1, where the nitrate content in lysimeter water from various depths is also low (Nilsson et al. 1988). With treatment N_2 nitrification is stronger in the tests, and nitrate concentrations considerably elevated in some but not all lysimeters. Treatment N_3 has so far not been included in this study, but earlier incubation tests (Tamm 1974) showed nitrification at that time. The conclusion which can be drawn at present is that the ecosystem has been able to absorb an increased nitrogen supply of 730 kg N ha^{-1} over a 20-year period without major changes in soil nitrogen metabolism (even if some leaching of the added nitrate ions may have occurred, as indicated by a slight acidification of mineral soil horizons (S. I. Nilsson, pers.comm.). The double amount of nitrogen has induced nitrification and thereby increased leaching. This process has also acidified the mineral soil and increased the aluminium concentrations in the soil water (Nilsson et al. 1988).

Similar conclusions can be drawn from the optimum nutrition experiments in young Scots pine (*Pinus sylvestris*) stands. Needle nitrogen levels have been increased in a similar way to those in the spruce experiment with similar nitrogen regimes (Fig. 4.11, Table 4.1). Changes in soil nitrogen turnover, in particular nitrification, were measured in laboratory incubations in 1974 (Popovic 1977). The differences from the controls in nitrate formation were small for

Table 4.3. Soil acidity at different nutrient regimes in experiments E 55 Norrliden, measured as pH in water suspension in June 1988. Arithmetic means of triplicate plots (without PK) or duplicate plots (with PK). For treatments, see Table 4.1. The pH levels given concern the humus layer (A_0) and the mineral soil at 10 – 20 cm depth (roughly the B horizon)

Soil layer	Nitrogen regime and nitrogen form									
	Urea, no PK					AN, no PK				
	N_0	N_1	N_2	N_3	LSD[a]	N_0	N_1	N_2	N_3	LSD[a]
A_0	4.48	4.80	4.86	4.81	0.21	4.54	4.81	4.94	4.94	0.29
B	5.63	5.60	4.76	4.38	0.32	5.60	5.59	4.57	4.38	0.34
	Urea with PK					AN with PK				
A_0	4.78	4.86	4.79	4.80	0.25	4.78	4.76	4.84	4.90	0.19
B	5.26	5.02	4.86	4.50	0.33	5.43	5.09	4.64	4.39	0.42

[a] LSD denotes the least significant difference at the 5% level.

regime N_1 with urea and for most plots with ammonium nitrate, while higher dosages of urea resulted in vigorous nitrification. In 1985 Högberg et al. (1986) measured the nitrate reductase activity in leaves of the grass *Deschampsia flexuosa* on the different treatments. They found elevated levels of the enzyme on all N_2 and N_3 plots, except on N_1 (urea) (Table 4.2), indicating that the grass roots had had access to nitrate. The increased enzyme level on N_1 (ammonium nitrate) plots is expected, as 30 kg ha^{-1} of ammonium nitrate nitrogen had been added earlier in the same season.

Soil studies (Table 4.3) have shown that both ammonium nitrate and urea applications have acidified the upper mineral soil. Except for some of the treatments with the lowest nitrogen fertilizer regime (N_1), the acidification of the mineral soil is statistically significant. In contrast, both nitrogen fertilizers and PK addition decrease the acidity of the humus layer. Another experiment at the Norrliden site allows comparison between the acidity changes from fertilization ($N_2P_2K_2$) and from application of 900 kg sulphuric acid ha^{-1} over a 6-year period. As seen from Fig. 4.12, both treatments have acidified the mineral soil of a typical podzol profile (an orthic podzol).

The acidification of the mineral soil by application of substances containing or forming anions of strong mineral acids (sulphate and nitrate) can be explained as being due to the downward percolation of mobile anions (Seip 1980), accompanied by both protons and basic cations. Together with the changes in acidity, as measured by pH or base saturation, there is a loss from the soil profile of basic cations, preferentially magnesium and calcium (Tamm and Popovic 1989).

The growth responses for fertilization have been much lower in Scots pine than in Norway spruce (Fig. 4.13). While the spruce diagram in Fig. 4.8 presented values for stemwood production in m^3 ha^{-1}, those for the pine stand

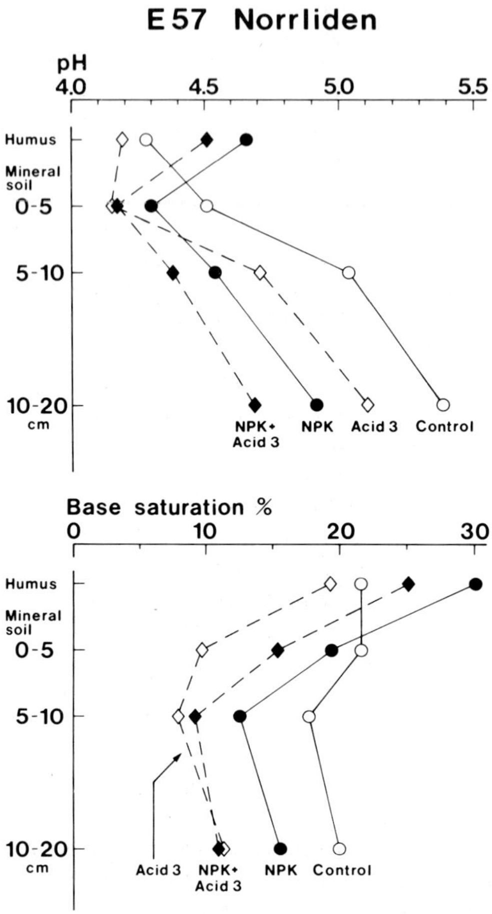

Fig. 4.12. Soil acidity (in water suspension, June 1985) and base saturation (at pH = 7) at different soil depth in Experiment E57 Norrliden. Treatment NPK is the same as $N_2P_2K_2$ in Table 4.1, lower part, and treatment Acid 3 means 900 kg H_2SO_4, in diluted form 1971–1976 (Tamm and Popovic 1989)

in Fig. 4.13 are relative values (controls = 100) of total stem production up to 1984, with the cubic metre values for the control plots at the bottom of the diagram. The data have been adjusted by analysis of covariance for differences in initial conditions of the stands (see legend to Fig. 4.13). Inspection of Fig. 4.13 shows that, despite the adjustment mentioned, there are some differences in levels of the curves for different treatments, making comparisons between urea (Fig. 4.13 A, B) and ammonium nitrate (C, D) difficult. Yet it is common to all diagrams that the initial reaction for nitrogen addition is strongly positive, and that the positive trend remains throughout the investigated period for the lowest nitrogen regime (N_1). For regime N_3 there is a levelling off in all four cases and even a tendency to growth decrease in later years. The curves for regime N_2 are intermediate, first increasing, then levelling off. Similar curves for relative basal area, measured on increment cores, have also been constructed. There are small differences in levels of the various treatments, as compared with Fig. 4.13, but the agreement in slopes for various

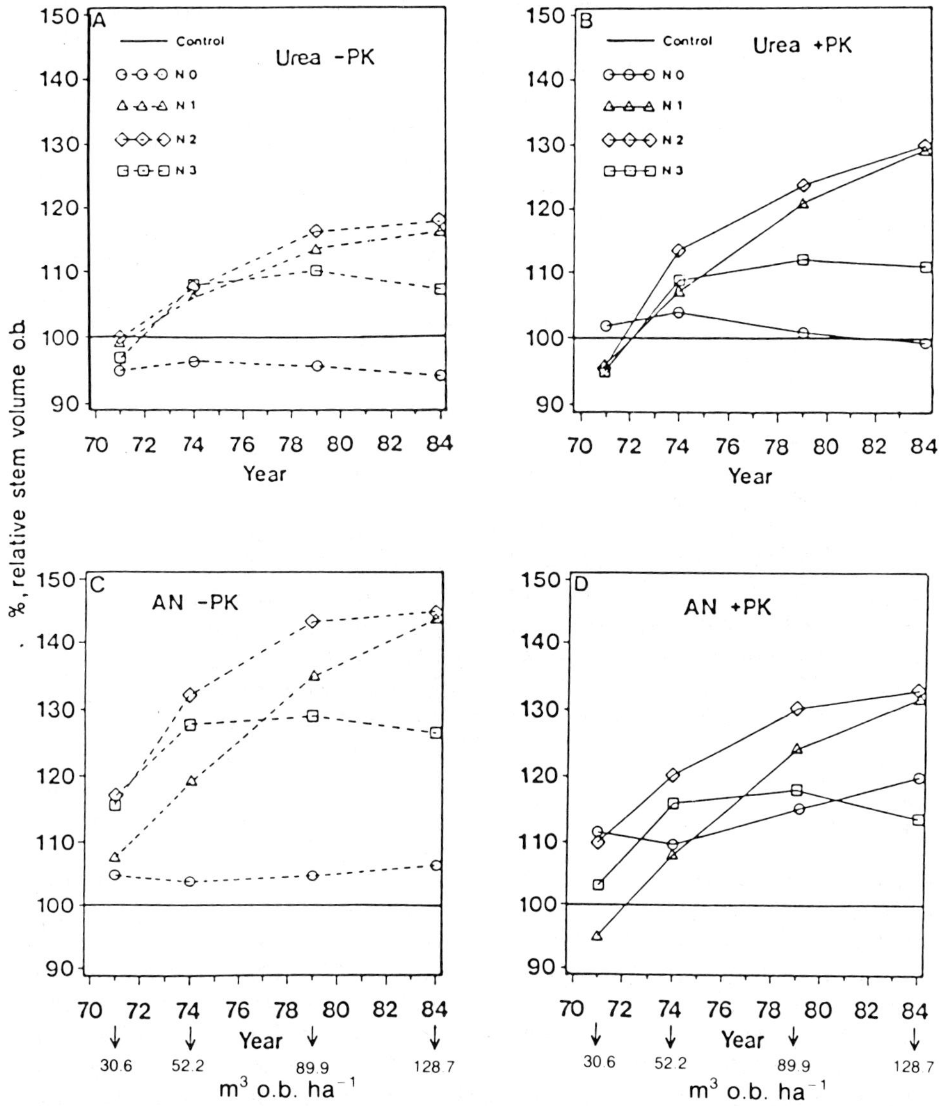

Fig. 4.13. Relative stem volumes over bark in different treatments of Experiment E55 Norrliden. The measured values have been adjusted for initial differences in "Björgung index", see Fig. 4.8. Mean of all six control plots = 100. NO in *left* diagrams denotes the mean of the three control plots in either the urea or the ammonium nitrate blocks. In the diagrams to the *right*, NO stands for PK. The values at *bottom*, below the *arrows* stand for the mean volumes in different years of the six control plots (Tamm 1989)

treatments is good, despite the very different way of measuring growth (Tamm et al. in prep.). It might also be mentioned that the tree shape changed with nitrogen regime. In pine, increased nitrogen supply stimulates branch and diameter growth more than height growth (e.g., Albrektson et al. 1977). The differences in spruce stem form associated with fertilization are less obvious than in pine (Mead and Tamm 1988).

As demonstrated in Table 4.3, for some horizons soil acidification increases with increasing nitrogen addition. With nitrogen regime N_1 urea, however, no significant acidification has occurred in any of the mineral soil horizons. There have not been direct studies of nitrate leaching in the Norrliden experiments, but analyses for ammonium and nitrate nitrogen on the same samples as in Table 4.3 show elevated levels of ammonium ions in most horizons after fertilization (Fig. 4.14). For nitrate there is also an increase with nitrogen regime, although most N_1 samples are low in nitrate in both humus layer and mineral soil. The samples were taken in June 1988, just before that year's fertilization, so most of the nitrate applied in 1987 would be expected to have been leached out at least from the humus layer (Melin and Nömmik 1988). Any larger amounts of nitrate on ammonium nitrate plots and all nitrate on urea plots must have been formed by nitrification.

Indirect evidence of leaching after fertilization at Norrliden can also be obtained from comparisons of the total nitrogen content in the soil (Fig. 4.14C). Data from a sampling in June 1988 indicate that plots with regime N_1 contain 294 kg N ha^{-1} more than controls in the humus layer and mineral soil down to 20 cm. This corresponds to about 46% of the amount applied in 1971–1987. The stand above ground also contains more nitrogen on N_1 than on control plots, roughly estimated to be about 16% of the amount applied. Smaller amounts would also be expected in the ground vegetation and in the soil below 20 cm. The amounts of fertilizer nitrogen recovered in soil and stand increase somewhat with increasing application, but expressed as percentages of added nitrogen the recovery in soil and stand is much lower at the nitrogen regimes N_2 and N_3 than at N_1. We may conclude that at the lowest nitrogen regime, at present 30 kg N ha^{-1} $year^{-1}$, most of the applied nitrogen is still retained. Some of the losses which have occurred may have taken place in the first years, when the rate of application was higher. It is also probable than some ammonia gas has been volatilized from the urea prills, a loss estimated to be between 0% and 15% in commercial forest fertilizations with urea, depending on weather conditions. During a period of 17 years, at least some years are likely to have favoured ammonia volatilization. With regimes N_2 and N_3, regardless of whether the nitrogen was given as ammonium nitrate or urea, large losses have occurred, as leaching, possibly denitrification and for urea volatilization.

As demonstrated in Table 4.3, soil acidification increases with increasing nitrogen addition, and so do nitrogen losses by leaching. In the future, tree growth may be even more disturbed than at present with nitrogen regime N_3, where signs of growth decline have already appeared (Fig. 4.13). With nitrogen regime N_2 the trees are still growing well, but as those with regime N_1 appear

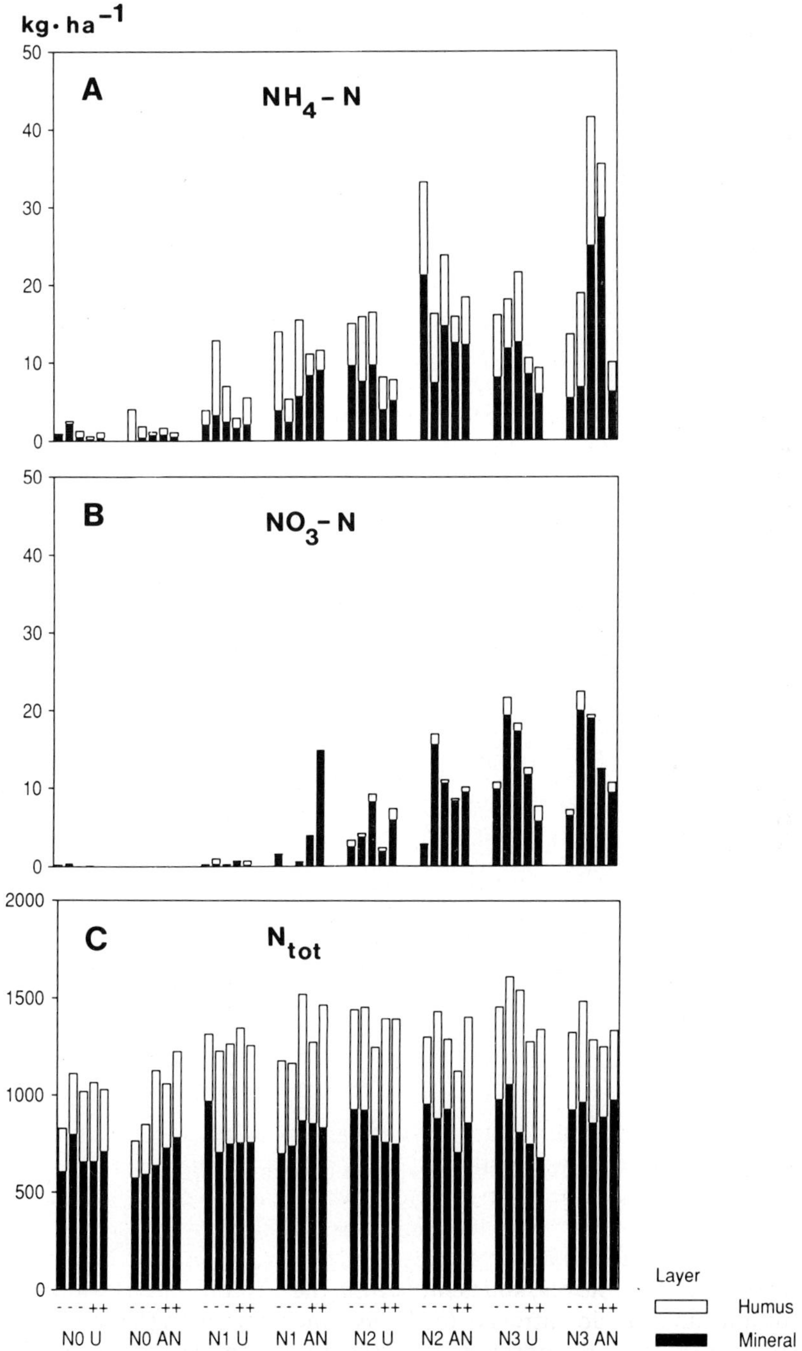

Fig. 4.14. Amounts in kg ha^{-1} of total and inorganic nitrogen in top soil (humus layer + 0 – 20 cm mineral soil) in Experiment E55 Norrliden, June 1988. **A** Ammonium nitrogen; **B** nitrate nitrogen; **C** total nitrogen. *Filled bars* mineral soil; *open bars* humus layer. Nitrogen regime (NO – N_3) and nitrogen form (*U* urea; *AN* ammonium nitrate) marked at bottom. *Plus* and *minus signs* at bottom denote with or without PK fertilizer. See further explanations in text

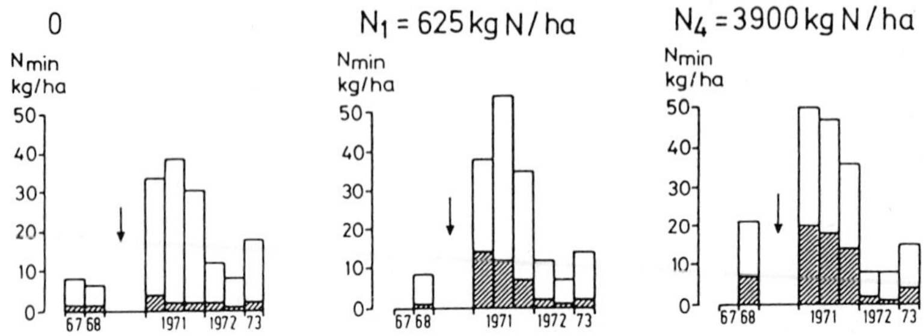

Fig. 4.15. Amounts of mineral nitrogen in top soil (0 – 20 cm) in some of the treatments in Experiment E1 Hökaberg, where a young stand of Norway spruce during the 13-year period 1957 – 1969 received annual additions of nitrogen ("Kalkammonsalpeter"), amounting to the totals noted at *top*. The site was abandoned farmland in middle south Sweden, planted in 1947 and clear-felled in 1979 (marked with *arrow*). *Empty bars* ammonium nitrogen; *solid bars* nitrate nitrogen (Tamm and Popovic 1974)

to catch up, there is considerable doubt about the long-term future of the stands on the N_2 plots. Nitrate leaching will increase when the tree growth and nutrient uptake is no longer stimulated, in accordance with the nitrogen saturation hypothesis.

Even if the ecosystem has so far retained the larger part of the nitrogen applied with regime N_1 there is a possibility that this nitrogen may remain more available than the native nitrogen, as more than half of it is contained in the humus layer. To a greater or smaller degree it may be released, if the tree stand is cut or removed in other ways (fire, storm-felling, etc.). We can expect elevated nitrate concentrations in the water leaving such sites as groundwater or surface runoff. Yet it is not clear whether sites enriched with nitrogen during decades of human influence will behave differently from naturally fertile sites, which have built up high nitrogen levels over centuries or longer. The experimental data available (Tamm and Popovic 1974; Fig. 4.15) indicate that intensive nitrification and consequently conditions favouring leaching occurred on an area cleared after 13 years of intensive fertilization (up to 3900 kg N ha^{-1} in total amount), but also that the nitrate concentrations in the soil and soil water returned to close to those from the unfertilized controls, as soon as the plots became revegetated.

4.5 Enrichment by Atmospheric Emissions from Modern Industrial and Agricultural Activities

A new enrichment factor besides chemical fertilizers is the increased atmospheric supply of nitrogen compounds to ecosystems in the industrialized world and also in intensively farmed areas. The ecological consequences will be discussed in Chapter 5. A more detailed discussion of the emissions of nitrogen compounds per se is beyond the scope of this work, but a few points should be made:

1. Nitrogen oxides, in particular NO, which is oxidized to NO_2 in the air, are produced in combustion processes, or as soon as air is heated to combustion temperatures. The largest sources for NOx are industrial combustion, motor vehicles, and house heating. Some ammonia may occur when the fuel contains organic nitrogen and combustion is incomplete. As previously mentioned in Section 4.1, there was always a background deposition of nitrate, mainly from lightning and wildfires, and of ammonium ions, probably mainly from alkaline soils and from decaying plant and animal residues, such as faeces, perhaps also to some extent from living plants (Farquhar et al. 1980, 1983).
2. Agriculture is a major source of both ammonia and nitrous oxide (N_2O). Ammonia evaporates from both ammonium-containing or ammonia-forming fertilizers (farmyard manure, urine, urea), preferentially but not exclusively from soils with high pH. Soil nitrate nitrogen is partly transformed to N_2O in the denitrification process but nitrification may also be a major source of N_2O (see Chap. 2). NO can also be formed in the same processes.
3. Free ammonia has a short residence time in the atmosphere, as has NO_2, but both gases may be transformed to aerosols with longer residence times (ammonium sulphate and nitric acid, the latter occurring also as a gas). The deposition of nitrogen in a rural area thus has several origins, one source area being the local neighbourhood (agriculture, motor traffic, combustion). Nitrogenous compounds are also transported long distances, and the balance between compounds of industrial origin and those coming from agricultural sources shifts towards a larger proportion of the former with increasing transport distance.
4. Both wet and dry deposition occurs. Wet deposition is simply the amounts coming in with rain, while the term dry deposition covers a number of different processes, absorption of gases, sedimentation of aerosol dry particles and droplets, filtering of particles blown through a forest canopy, etc. Wet deposition is relatively evenly distributed, related to rainfall. Compared with wet deposition, the dry deposition depends more on surface area, geometry, and for gases, the chemistry of the absorbing structure. Forests expose much larger surfaces to the wind than open fields and can therefore be expected to absorb more of gaseous ammonia, nitrogen dioxide, and gaseous nitric acid, all with a high affinity to plant cell surfaces. For dry deposition of aerosols, such as ammonium sulphate and droplets containing nitric acid, surface geometry and structure, especially its resistance to wind, plays the largest role. Forest edges get higher deposition of both gases and aerosols than the interior of a forest or an open field (Grennfelt and Hultberg 1986) for rather simple physical reasons. The same is true to a certain extent also for wet deposition, especially when rain drops are small. So-called fog drip, common in mountains, is a transitional form between wet and dry deposition, often with high concentrations of pollutants.

The ecological consequences of nitrogen emissions will be further discussed in the next section, so we will stop here with the statement that the recent nitrogen enrichment of land in Europe and other densely populated and industrialized parts of the world is a major environmental change. In agricultural areas, enrichment by fertilization and deposition may be compensated for to a variable degree by removal due to harvests. There are good indications that agroecosystems in these areas generally have a higher nitrogen status now than a century ago, and that losses of nitrogen from the systems both to the atmosphere and by runoff have increased in the last decades. The causes are of course not nitrogen deposition alone, but even more increases in nitrogen fertilization. In forest and other natural and seminatural vegetation less intensively cropped, the ecosystems are likely to develop towards a nitrogen-saturated situation. The rate of this change is extremely variable, depending not only on the rate of enrichment by deposition but also on the character of the ecosystem, especially its capacity for nitrification and denitrification, and its supply of other plant nutrients. The acidification due to deposition of acids and acid-forming substances may also interact with the nitrogen cycle by enhancing or retarding proton-producing or proton-removing processes such as nitrification and denitrification, as discussed in Sects. 2.2.2 and 2.2.3.

5 Consequences of Increased Nitrogen Supply to Forests and Other Natural and Seminatural Terrestrial Ecosystems

5.1 Changes in Vegetation and Fauna

5.1.1 Forest Vegetation

As described in Chapter 3, terrestrial ecosystems are often nitrogen-limited, either because of low total amounts or because of low availability. Even in cases where the total primary production is not severely hampered by lack of nitrogen, some ecosystem components may suffer from nitrogen deficiency. The addition of nitrogen, either by fertilization or by increased atmospheric deposition, is then likely to increase growth of some organisms, including at least some of the primary producers. An adequate name for this effect is eutrophication. The term is well-established for lakes receiving an increased supply of nutrients but is equally useful for terrestrial ecosystems, even if the limiting nutrient differs between different ecosystems. There are scores of publications on positive responses in tree growth after nitrogen application and also a few reports on differential responses in different tree species. As most foresters prefer to experiment with single-species stands (monocultures or naturally established) to minimize experimental error, species comparisons are relatively rare, except for nursery conditions and young plantations. One of the best experimental studies is still that by Mitchell and Chandler (1939) in the northeastern United States. They carried out nitrogen fertilizer experiments in mixed hardwood stands and measured the changes in foliar nitrogen concentrations and in stem radial growth. Their results allowed them to classify a number of tree species as nitrogen-demanding, nitrogen-tolerant, or intermediate (Fig. 5.1). The nitrogen-tolerant species, like trembling aspen and red oak, grew comparatively well on sites with low nitrogen availability but did not increase growth after fertilization as much as the nitrogen-demanding species such as white ash and yellow poplar. Similar species differences also occur between Scots pine and Norway spruce (Tamm 1985; Sect. 4.4), even if they are difficult to deduce from conventional trials in single-species stands. Chapin et al. (1986) discuss the influence of both site fertility and stand age on the responsiveness to added fertilizer in some boreal tree species.

A safe conclusion on the basis of what has been said above is that changes in atmospheric nitrogen deposition are bound to change the forest succession pattern in all cases where there is a competition between species with differences in nitrogen demand (cf. Chapin et al. 1986). Unfortunately, natural suc-

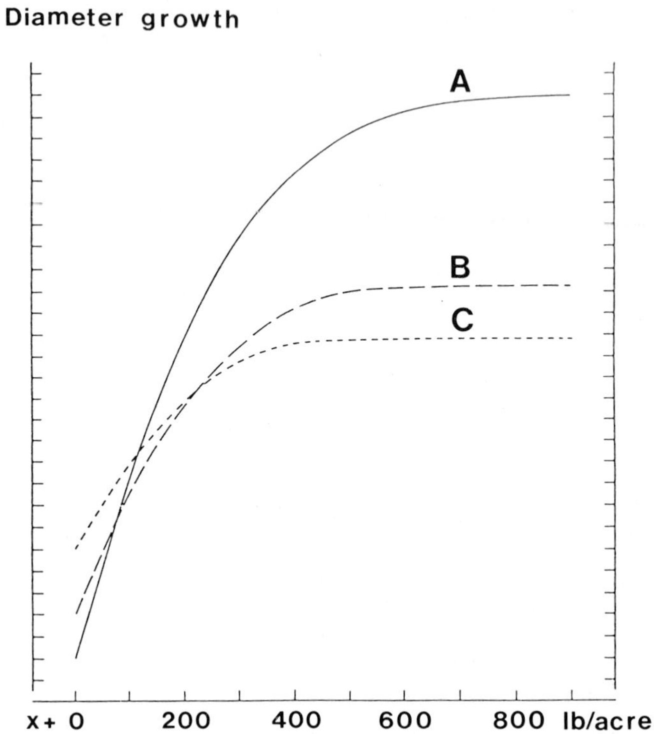

Fig. 5.1. The relative growth response to increasing nitrogen supplies of **A** nitrogen-demanding, **B** intermediate and **C** nitrogen-tolerant tree species. The scale on the y-axis shows the annual radial increment and that on the x-axis the amount of nitrogen applied (in addition to the unknown amount x available on control plots on the poorest site studied). Trembling aspen, red and white oak, and red maple were among the species classified as nitrogen-tolerant, while white ash, yellow poplar, and basswood were considered demanding. Sugar maple and beech belonged to the intermediate group. The classification is based upon a number of fertilizer experiments in mixed hardwood stands in the northeastern United States (Mitchell and Chandler 1939)

cession is seldom allowed to develop undisturbed in the regions where nitrogen deposition is increased by human activities. In addition, eutrophication by nitrogen input is likely to be confounded with soil acidification and so-called novel forest damages, also caused by air pollutants. The long lifetime of individual trees makes it difficult to study consequences of a factor which in most areas has been of importance for less than 50 years, a period which has also witnessed important changes in land use (cessation of forest grazing and litter removal, etc.). Some simulation models for forest stand development and ecosystem succession (Aber and Melillo 1982; Kimmins and Scoullar 1984; see also Dyck et al. 1986) include plant nutrient relations (mostly as limiting factors), but experimental validation is, for natural reasons, difficult, with virtually no fully adequate long-term observations of changes in nutrient availability.

If, therefore, field studies are rare or lacking in tree succession in response to increasing nitrogen supply, there are many reports on changes in forest undergrowth after fertilizer applications (e.g., Gerhardt and Kellner 1986; Dirkse and van Dobben 1989). Tilman's data from "oak savannah" were mentioned earlier (Sect. 4.4). As most forest experiments are designed to answer practical questions, it may in some cases be difficult to separate nitrogen effects from those of other treatments (e.g., soil preparation), or the results may be representative for one phase only in stand development (e.g., the regeneration phase).

Regional comparisons of forest undergrowth in oakwoods in southern Sweden show differences in frequencies of nitrophilous species (*sensu* Ellenberg 1979), correlated with differences in nitrogen deposition (Tyler 1987). Data on vegetational changes on forest plots over the last few decades support the hypothesis of a causal relationship (Falkengren-Grerup 1986; Falkengren-Grerup and Eriksson in press).

Some general conclusions can be drawn. If a forest stand responds positively to nitrogen, and the canopy density permits a field layer vegetation, the components of this vegetation are likely to respond positively to nitrogen. There are physiological reasons to believe that shaded plants in the forests are at a disadvantage when nitrogen is scarce (Sect. 2.1). In addition to uptake costs, utilization of ammonium nitrogen requires the presence of organic acids. Utilization of nitrate has even higher energy costs, particularly if the nitrate is reduced in the roots. Typical plants of strongly nitrogen-limited forests are ericaceous species, which, on their normal sites, react positively to supply of ammonium nitrogen. In somewhat richer forest types, with nitrification, several plant species are "nitrate plants" (Hesselman 1917; see also Sect. 2.1) in the sense that they accumulate nitrate in their leaves, due to low nitrate reductase activity in the roots (e.g., *Lamium galeobdolon*, *Stachys* sp.). In both these cases there is often a positive growth response in the field layer to nitrogen fertilization. Often some species respond more than others. On poor sites *Deschampsia flexuosa* often occurs in scattered stunted specimens and may become dominant after nitrogen fertilization (Fig. 2.5). Yet the effects of increased nutrients supply may soon be suppressed by increased shading from a denser canopy. In a case described by Tamm (1974), annual fertilization for 6 years of a young spruce stand on a site low in available nitrogen resulted in the complete replacement of the original dwarf shrubs with luxuriant *Rubus ideaeus* (a nitrate plant) on plots with high nitrogen levels (Fig. 4.5). *Epilobium angustifolium*, another nitrate plant, also increased. Yet the spruce also responded, and 5 years later most field layer vegetation had disappeared beneath the now very dense canopy of the fertilized plots.

5.1.2 Grasslands, Heathlands, and Wetlands

In Sect. 4.4, we described some of the ecosystem changes occurring in grasslands with increased use of nitrogenous fertilizers. These changes are of course not specific for enrichment of nitrogen by fertilization, but can also be expected as consequences of increased nitrogen deposition. There is, however, reason to discuss in some more detail the processes occurring and possible shifts from one type of ecosystem to another as a consequence of nitrogen enrichment.

We have already mentioned the negative relations between the amounts of biomass and litter produced by the field layer, on the one hand, and the species diversity, on the other (Fig. 4.2; see also Grime 1979). At high nitrogen supply and low rates of biomass losses only plants with large allocations to height growth will be competitive (Tilman 1988). The development can be observed on abandoned meadows and pastures when the removal of nutrients by haymaking or grazing is discontinued. Tall grasses together with some tall herbs such as *Anthriscus silvestris* become dominant on well-drained sites and allow few other species to coexist. Sites with drainage problems are often taken over almost completely by *Filipendula ulmaria*. The species mentioned are characteristic for this development in northern Europe, and other species play a corresponding role in other regions. In such dense herb or herb-grass vegetation even tree species have difficulties colonizing, except by root suckers (alders). An adequate name for this process is self-eutrophication, and the only way to preserve the original species diversity is by regular removal of plant biomass. This is done in some nature reserves, but, because of the costs, only on limited areas, and not always in the same way as in the old management.

The situation is less clear for temperate wetlands and also for some west European heathlands. Here phosphorus is often more limiting than nitrogen, and also potassium may be in low supply, as plants on drained peatlands often display symptoms of potassium deficiency. Yet it would be surprising if not an increase in nitrogen deposition from the few kilograms per hectare and year which have been estimated for preindustrial time to 20 or more kilograms in central Europe and southernmost Scandinavia would change the floristic composition of ombrotrophic bogs. There is a need for systematic studies of such vegetation changes, mainly within nature reserves, where no changes in land use occur. Vegetation types such as ombrotrophic bogs and heathlands on poor soils depend far more on the chemical composition of the rain than on nutrients from the subsoil and would be particularly susceptible. It has already been reported that protected Dutch heathlands are losing their *Calluna* and are becoming more grassy (Roelofs et al. 1987). The disappearance of *Sphagnum* moss from the moorlands of the Pennines in northern England have been studied by Lee (1986) and Lee and Woodin (1988), and explained at least partly as being an effect of NO_2 deposition.

A discussion of vegetational changes associated with eutrophication must also consider epiphytic lichens. Lichens are known to be sensitive to air pollution, particularly to sulphur dioxide and, because of the different sensitivity of different species, they have been used to map the air pollution situation (Skye 1968; Barkman 1969; Hawksworth and Rose 1970). Some species, however, e.g., *Xanthoria parietina*, are favoured by a rich supply of nitrogen and occur, e.g., on bird cliffs and on trees exposed to dust and perhaps gases from arable land. They might be favoured by at least a moderate eutrophication, while there are reasons to suspect negative effects in many other species.

5.1.3 Fauna

While the title of Chapter 5 deals with ecosystem changes related to changes in nitrogen availability, Sects. 5.1.1 and 5.1.2 have only dealt with vegetation types. The reason for this deviation from a fully logic structure are purely practical: the causal relationships behind observed changes in animal populations are much less clear than in the case of plants. Animals represent higher levels in the food web than green plants, and they are thus directly or indirectly affected by all factors influencing their food and habitat, at the same time as they interact with other animals and with humans. In addition, many animals move regularly between different ecosystems, nearby or far away (migratory birds).

However, animals are equally as essential constituents of natural ecosystems as plants. Their reaction to environmental changes such as the increase in nitrogen deposition must therefore be discussed, on the basis of available observation together with common ecological knowledge. It is to be hoped that future attempts to synthesize effects of nitrogen enrichment on animal populations will be founded on more facts than those presently available to the author. The feedbacks on ecosystem functions of changes in animal populations are even less well understood than the causes of the variations in herbivore and predator populations. The discussion here will deal mainly with vertebrate fauna, as some insect-plant relationships will be discussed in Sect. 5.3.6.

H. Ellenberg Jr. (in press) first reviews some of the landscape changes connected with eutrophication due to increased nitrogen supply (changes that are discussed in Chap. 4 and the first part of Chap. 5) and concludes that vertebrate species favoured by the development up to a certain threshold are species such as: 1) deer; 2) wild boar; 3) wintering arctic geese and swans; 4) wood pigeons (not only wintering), and 5) some duck species (mainly wintering).

Declining under the same conditions are species such as: 1) quail; 2) partridge; 3) rabbit; 4) hare; and 5) many bird species.

The causal relationships are reasonably well-proven in some cases only, e.g., for roe deer, which prefers nitrogen-demanding food plants (H. Ellenberg Jr. 1988). However, there are several feedbacks in the system: strong browsing on preferred species decreases their coverage and emission damages on the tree stand may open the canopy and provide more plant growth in the height range accessible to deer.

The numbers of moose in Scandinavia have increased considerably during the last 50 years, which is considered to be mainly due to changes in forest management, with more clear-felled areas and reforestation providing good moose browsing. As it is known that the moose prefers nitrogen-fertilized pine shoots to unfertilized ones (Brantseg 1966), food quality may also play a role. Another report of preferences for high-nitrogen food concerns African ungulates (Coe 1983), but protein content was only one of several factors affecting food selection. Tamm (1974) reported that moose appeared to find spruce shoots fertilized with both nitrogen and phosphorus most palatable, while voles eating the bark of young spruce trees only reacted to nitrogen. However, the shelter provided by the luxuriant field layer vegetation on nitrogen-fertilized plots might have been more important for the rodents than the chemical composition. Further discussion on the importance of the carbon/nitrogen ratio of vertebrate food is given by Bryant et al. (1983).

In the case of some of the waterfowl species favoured by eutrophication there is also reason to believe that the food available for them has increased by eutrophication of their preferred sites, shallow waters, and adjacent land. According to Rüger et al. (1986), the duck species *Anas crecca*, *A. clupeata*, and *A. strepera* increased in Central Europe during the period 1967 to 1983.

Mammals and birds decreasing with eutrophication, according to H. Ellenberg Jr. (in press), are those preferring a relatively open vegetation, with a diverse vegetation of short or moderately tall herbs and grasses, and consequently a good supply of seeds and various types of insects, available during a longer part of the season than on sites where one or a few tall species dominate the field layer. The better visibility of the environment and thereby better possibility of detecting approaching enemies is also an important factor for a number of species preferring more open areas.

Lists similar to those for mammals and birds can be made up for invertebrates in relation to changes in the chemical environment. Insects are often much more specifically restricted to certain host plants than grazing vertebrates, and a list of endangered plant species thus automatically provides examples of endangered insect species. The recent decline in Scandinavia of the butterflies *Parnassias apollo* (with caterpillars on *Sedum*) and *Parnassias mnemosyne* (caterpillars on *Corydalis*) may have something to do with adverse changes in the sites where the host plants occur, even if the host plants are not immediately threatened. Insects and other invertebrates depend more directly on microclimate than most vertebrates and birds, and eutrophication certainly changes microclimate, usually from a more continental type to a more oceanic one (H. Ellenberg Jr. 1985).

5.1.4 Discussion

An increase in nitrogen deposition means a radical change in competitional advantages between species (cf. Tilman 1988). Sensitive ecosystem types occur over a considerable part of the earth's land surface. The effects would be smallest in areas where primary production is controlled by lack of moisture or by deficiency in some other plant nutrient such as phosphorus. In Europe most forests are managed, so the species composition of the dominating tree layer is human-controlled, but other vegetation components can be expected to change in the direction of less symbiotic nitrogen fixers, less nitrogen-tolerant species (*sensu* Mitchell and Chandler, Sect. 5.1.1) such as most dwarf shrubs, as well as many herbs, mosses, and lichens, and in more of nitrogen-demanding plants such as those common in richer forest types. Where light allows, species characteristic for disturbed sites or early stages of secondary succession will invade, like fireweed, raspberry, and ruderals (cf. Ellenberg Sr. 1979; Persson 1981; Wittig et al. 1985; Ellenberg H. Jr. 1985; Tyler 1987).

There is an urgent need for reliable indicators of nitrogen status of trees and other vegetation, even if there is the well-developed Ellenberg system of indicator plants (Ellenberg Sr. 1979) which works well in Central Europe. However, indicator plant systems give only semiquantitative information on nitrogen status and, in addition, some established plant communities may have a certain inertia, meaning that changes in vegetation do not show up in full until there has been some disturbance.

Soil analyses, including incubation experiments, have not been very successful in characterizing nitrogen availability in natural and seminatural ecosystems, for reasons discussed earlier (Sect. 2.1, see also Keeney 1980).

We have already mentioned in the introduction the possibility of using diagnostic plant analysis as an indicator system (Fig. 1.4). However, concentrations of total nitrogen in foliage or other plant organs are probably not the best indicator of excess nitrogen, i.a., because of its rapid variation during the growing period (Tamm 1964; van den Driessche 1974). It is well known that conifers often accumulate excess nitrogen as arginine (while many other plants accumulate glutamine or asparagine under similar conditions). Results from the Swedish optimum nutrition experiments (Fig. 5.2) indicate that arginine accumulates in spruce needles at a disturbed balance between phosphorus supply

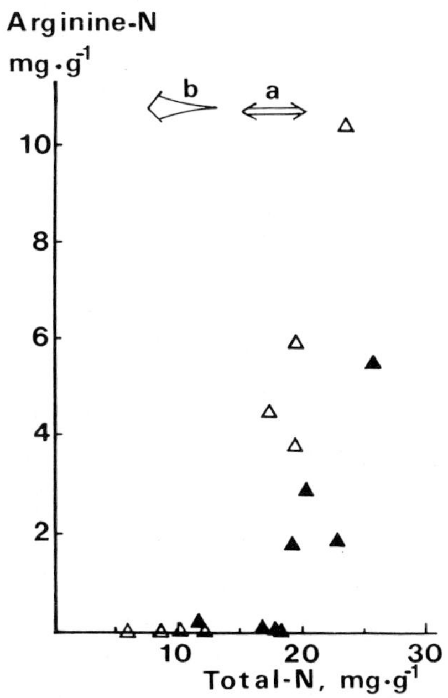

Fig. 5.2. Arginine nitrogen in spruce needles from various treatments and samplings in the Stråsan experiment, plotted against total nitrogen, mg g^{-1} dry wt. At total nitrogen concentrations below 15 mg g^{-1}, nitrogen is the main limiting factor and no arginine accumulates. At higher total nitrogen concentrations, arginine accumulates when nutrient conditions are unbalanced, either because of the lack of phosphorus or of excessive nitrogen supply. *Filled triangles* current needles; *open triangles* 1-year-old needles (Aronsson, unpubl. data)

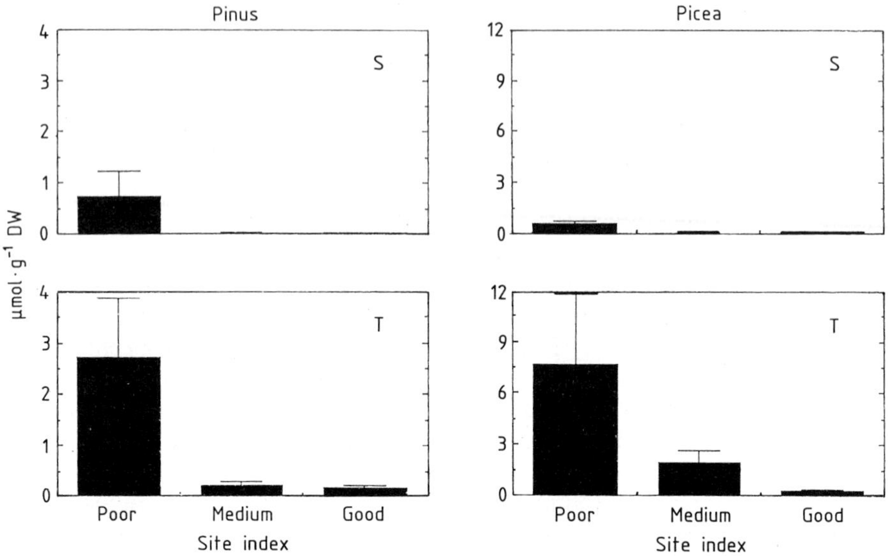

Fig. 5.3. Variation in arginine concentration in 1-year-old needles of *Pinus sylvestris* and *Picea abies* with site quality in two areas with very different atmospheric depositions of nitrogen: *S* Svartberget Experimental Forest, N Sweden, low deposition, and *T* Tönnersjöheden Experimental Forest, SW Sweden, high deposition. Sampling in June 1988, during the shoot elongation period. In all cases there are higher arginine concentrations on the poor sites than on the better ones, suggesting less ability of the more slow-growing trees to make use of their available nitrogen (A. B. Edfast, T. Näsholm, A. Ericsson, unpubl.)

in relation to nitrogen supply. As long as the nutrition is balanced, arginine concentrations remain low, also at nitrogen concentrations up to almost 2% dry weight, which are at or above optimum (cf. Figs. 4.7, 4.8). Still higher nitrogen concentrations *or* unbalanced nutrition leads to arginine accumulation.

The possibility of using arginine as an indicator of unbalanced nutrient supply is also suggested by the data in Fig. 5.3. The needles sampled for this investigation were collected in June during the shoot elongation period from two areas with large differences in atmospheric nitrogen deposition, in both cases from stands of different site classes. It is evident that possible accumulations of arginine in needles of pine and spruce have been depleted on good sites in both areas (where intensive growth during the elongation period creates effective sinks for nitrogen), while there is a tendency that the hampered growth on poor sites (due to water stress or lack of mineral nutrients) leaves some stored arginine unexploited during this period.

Treeless vegetation such as heathlands and grasslands would be expected to change considerably with increasing nitrogen level, with shifts in species composition, according to the Ellenberg system. Legumes and other symbiotic nitrogen fixers would stop fixation and thereby lose their competitional advantage. A few tall herbs and grasses would become dominant, and the species diversity would decrease.

Dry sites, where the shallowness of the soil and/or strong seasonal drought prevents tall herbs or grasses from forming dense layers, may still allow some weaker (but drought-resistant) competitors to survive. When drought is exceptional, woody plants, if occurring, may die, which favours invasion of ephemerals such as *Poa annua*, when the rain comes. There is a risk that an increased nitrogen level results in a more permanent ruderal vegetation instead of the original low but species-rich vegetation. Such a development has happened on the "alvar" (a kind of karst vegetation) on the island of Öland, in south Sweden, under the influence of heavy sheep grazing (Rosén 1982).

Sites characterized by low amounts of available phosphorus would of course be expected to change less with increases in nitrogen deposition than nitrogen-limited ecosystems. It is then probably less important what the causes are of the low phosphorus availability. Phosphorus limitation may be a natural phenomenon of ombrotrophic bogs or of iron-rich phosphorus-fixing soils in temperate and tropical regions, but may also be due to human removal, as on land long used for hay-making or grazing. Still, there are probably few vegetation types, at least in temperate and reasonably humid parts of Europe, which would not show any changes with a strongly increasing load of nitrogen. And while the change might mean an increase in primary production, there is probably a decrease in floristic diversity, as most existing species are adapted to well-defined and mostly comparatively stable ecological niches, while the new situation created by nitrogen enrichment can be expected to disfavour many established species but suit ubiquitous ruderals (cf. Sect. 4.4; Tilman 1986, 1987, 1988).

5.2 Soil Chemical and Biological Changes, Including Nitrogen Losses to Groundwater and Atmosphere

5.2.1 Factors Affecting Biological Activity in Soils with Differences in Nitrogen Supply

Biological soil processes and their relations to the nitrogen supply were discussed earlier (Sect. 2.2). While the production of green plants is often limited by available nitrogen, with energy supply in the form of light as a factor in relative abundance at least for upper strata of vegetation, heterotrophic organisms in the soil depend mainly on energy supply. Plant litter and exudates constitute the main sources of energy-rich organic compounds. Yet other nutrients are by no means without importance, as decomposers and predators on them also need mineral nutrients, with nitrogen as the quantitatively most important one. Other factors influencing the decomposer organisms and hence the decomposition rate are of course temperature, moisture, and soil acidity. As different organisms often have different activity ranges or preferences with respect to both chemical and physical conditions, changes in one factor may lead to more complicated reactions than just a change in rate of a specific process. We have, for instance, mentioned earlier the possibility of a discoupling between nitrification and root uptake of nitrate (Ulrich 1983), with further consequences for many soil processes.

The balance between different pathways for the decomposition – with some oversimplification, the difference between mull and mor as decomposition systems – is sensitive to various disturbances, such as soil acidification, change of litter quality, and availability of nitrogen. Changes in soil decomposition type will also influence the soil physical structure and the spatial distribution of mineral nutrients as well as of harmful substances such as aluminium. There is evidence that root distribution and root efficiency are adversely affected in acidified soils, where, however, most of the anthropogenic soil acidification so far has been ascribed to sulphur compounds. Most mycorrhizal fungi prefer ammonia (and possibly low-molecular organic nitrogen) to nitrate (Alexander 1983). Changes in pH level and in the ratio ammonium to nitrate ions will shift the balance between different fungal species and possibly between mycorrhizal and non-mycorrhizal nutrient uptake. Presence of ammonium ions in large amounts in acid soil may also affect the ability of existing mycorrhiza to take up metal cations (Boxman et al. 1986).

Further changes also mentioned earlier are changes in ecosystem output, with respect to both drainage water and the atmosphere. Increased amounts of nitrate in drainage water inevitably increase losses of cations such as calcium, magnesium, and potassium, and lead to increased transport of protons and/or aluminium to aquatic ecosystems. The amounts of gaseous losses of nitrogen also increase with higher internal levels, and the proportion of N_2O in losses by denitrification is likely to increase with increasing acidity.

5.2.2 Consequences of Increased Nitrogen Supply

It is not easy to predict quantitatively the consequences for the decomposition process of increased nitrogen deposition, with or without an accompanying pollution-caused soil acidification. The first reaction in mull soil is the simplest case: there will be more nitrate present in the profile and therefore more leaching of nitrate (with accompanying cations) to the groundwater. The denitrification rate may increase with increasing nitrate concentrations, depending on the occurrence of microsites in the soil with limited oxygen concentrations. Nitrification is an acidifying process, and deposition of reduced nitrogen subsequently transformed to nitrate thus contributes to soil acidification together with sulphur compounds and the protons associated with deposited nitrate ions. Although the denitrification consumes protons, this process will not reverse the acidification unless there is a quantitative denitrification of all nitrate originating from deposited ammonium nitrogen. The acidified soil may become inhabitable for earthworms and change towards mor humus and podzolization. Increasing soil acidity should, according to Sect. 2.2.2, tend to decrease nitrification, but there is no reason to believe that this retardation will be a rapid process unless the whole soil profile becomes extremely acid. Leaching losses can thus be expected to continue even in an acidified but nitrogen-rich soil.

It is possible that chemical or biological denitrification in deeper soil layers may help to remove some of the nitrate before the water reaches wells for human consumption, as seems to happen in some cases where the source of nitrate is agricultural fertilization (Lind 1979). In a mor type of humus layer the forest floor is already very acid and has a low base saturation. The decomposer organisms attack litter material with high C/N ratio, such as conifer needles, and during one phase in the decomposition the rate is apparently limited by access to nitrogen (and perhaps phosphorus). As there can be a net increase in nitrogen during certain stages of the decomposition process (in litter with a high C/N ratio), there seems to be a transfer to the upper litter from somewhat lower and more decomposed material by fungal hyphae (Sect. 2.3.1). This mechanism seems to be an adaptation to an environment with low amounts of available nitrogen and leads to small losses of nitrogen under undisturbed conditions. Even in cases where there is no net accumulation of nitrogen in the beginning of the decomposition process, there are no or small losses of nitrogen in the early stages of breakdown, and in later stages the losses will be proportional to mass losses, which in turn depend on lignin concentrations (Berg 1986). In some cases the lignin concentrations in the litter increase with increasing nitrogen availability and, in addition, lignin seems to decompose slower in nitrogen-rich litter (Berg 1986). On the whole, the role of soil animals seems to be to enhance decomposition and thereby mineralization (with the conditions in a rich mull with digging earthworms being an extreme case).

The first consequences of increased nitrogen supply to a nitrogen-limited forest of conifers or nitrogen-tolerant hardwoods on poor soils will be changes in litter quantity and quality. Vitousek (1982) has demonstrated what seems to be a universal relationship between the amounts of nitrogen in the litterfall and the carbon/nitrogen ratio of the litter (Fig. 5.4). Larger amounts of litter, even with unchanged C/N ratio, would mean larger amounts of nitrogen cycling, but the additional lowering of the C/N ratio is likely to promote nitrification (Fig. 5.5, from Kriebitzsch 1978). The concentrations of mineral nitrogen in the soil will increase, leading, i.a., to the changes in botanical composition earlier described. Because of the existence of negative feedbacks in the system (some retardation of the decomposition of nitrogen-rich litter, which in addition to more nitrogen may contain more lignin, possibly also formation of more resistant humus fractions), there will be a certain accumulation of nitrogen in the soil, in various phases of decomposition, in the A horizon.

Kriebitzsch (1978) distinguishes (in agreement with Zöttl 1960) between four different patterns of nitrogen mineralization (Fig. 5.6). His data are based upon samples from northern Germany, but the results seem to have a rather general applicability. It can be concluded that some soils (A) are very resistant to nitrification, while others have a retarded or incomplete nitrification (B and C). In some soils all mineralized nitrogen is rapidly transformed to nitrate on incubation (D). It is a natural conclusion that increases in nitrogen deposition will move the system along the trend from A to D. Yet there are reports from the Netherlands indicating that some ecosystems with very high ammonium

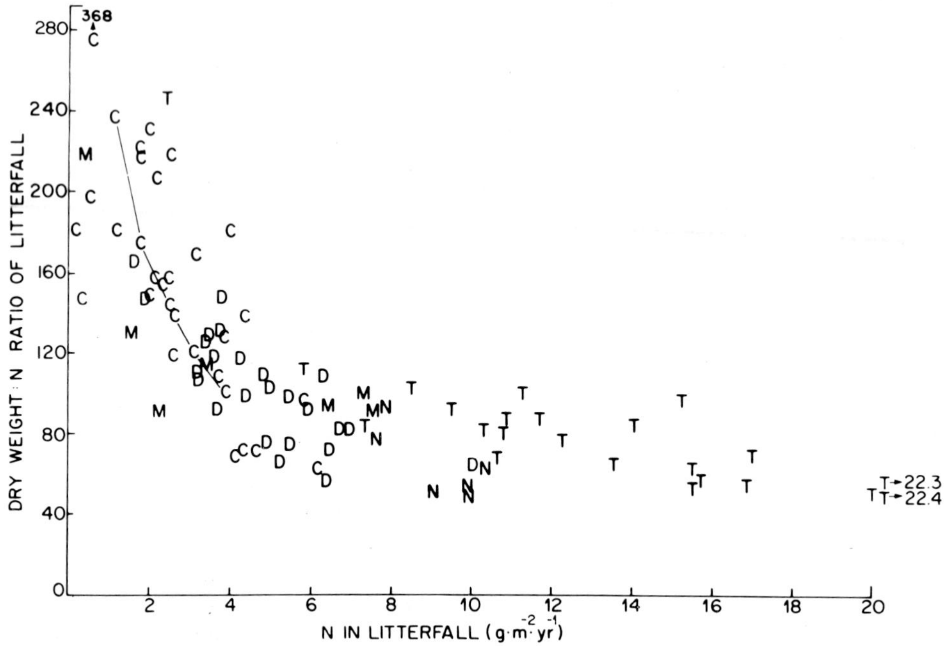

Fig. 5.4. The relationship between the amount of nitrogen in litterfall and the dry mass to nitrogen ratio of that litterfall. Different ecosystem types marked with different letters: *T* tropical; *D* deciduous; *N* temperate symbiotic nitrogen fixing; *C* coniferous; *M* Mediterranean forests (After Vitousek 1982)

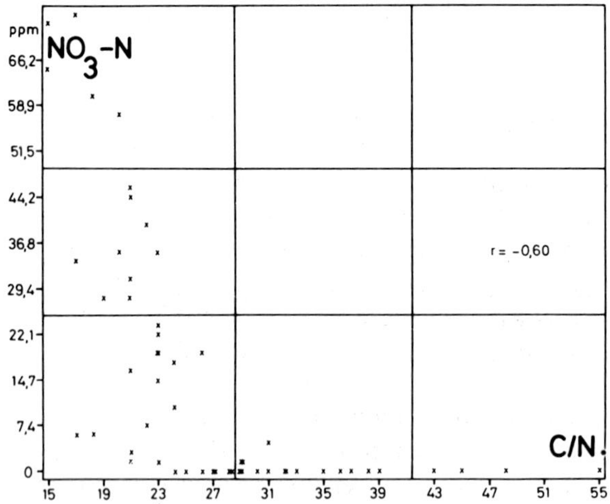

Fig. 5.5. Nitrate production in incubation experiments in humus samples plotted against the C : N ratio in the same horizon. Sites from northwestern Germany (Kriebitzsch 1978)

input may also have impeded nitrification (Boxman et al. 1988). The result is then both high internal ammonium concentrations and loss of ammonium nitrogen with runoff.

Most investigations on nitrogen turnover in forest soils (including Kriebitzsch 1978) are restricted to the forest floor and the top mineral soil.

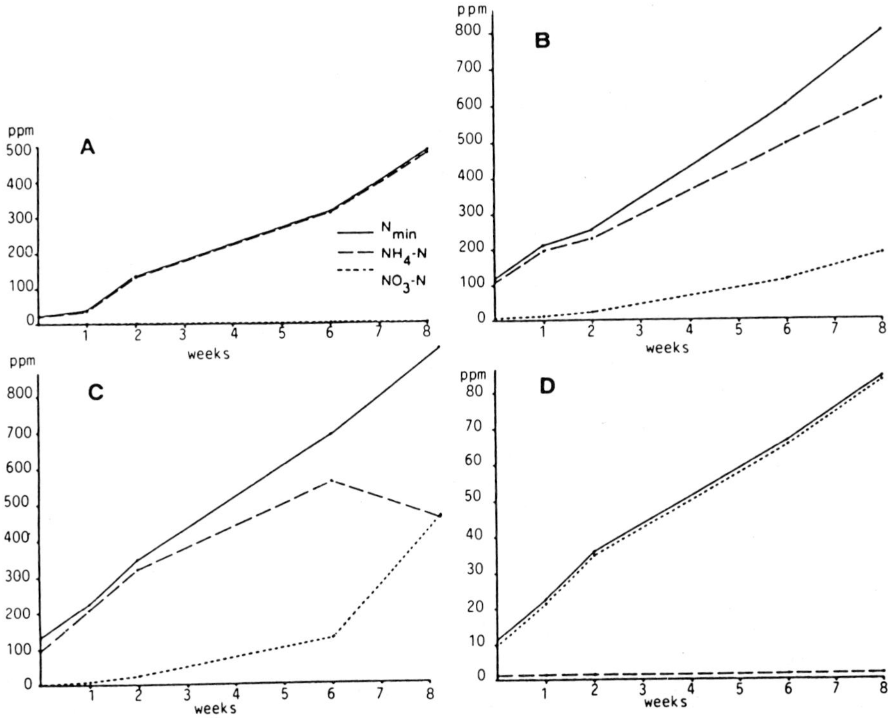

Fig. 5.6. Nitrogen production pattern during incubation in the four groups of forest soils as defined by Zöttl (1960): **A** Non-nitrifying soils; **B** low-nitrifying soils; **C** nitrifying soils; **D** total nitrifying soils (After Kriebitzsch 1978)

What happens in the B horizon upon increased nitrogen supply is less clear. It normally contains a large share of the total amount of nitrogen in a boreal or temperate forest (Tamm and Holmen 1967). Most of this nitrogen is bound in organic matter precipitated together with iron and aluminium hydroxides in a form apparently fairly resistant to decomposition, judging from the radiocarbon age of the material (Tamm and Holmen 1967). It seems likely that a slow increase in nitrogen content will happen also here, as the C/N ratio normally is higher than in the organic matter from more productive ecosystems. On the other hand, there may be less resistance to nitrification there than in the A horizon, and the "acid flush" described by Ulrich (1983, cf. Sect. 2.2.2) is a phenomenon taking place at least partly in the mineral soil.

The first phase of reaction to increased nitrogen deposition to a nitrogen-limited ecosystem will most likely be increased accumulation of nitrogen in both biomass (increased by the growth-stimulating effect of the nitrogen) and soil. No dramatic effects on soil processes will be observed in this first phase, but apart from the increase in tree growth, there will be changes in lesser vegetation (Sect. 4.4) and microflora (including macrofungi). According to current theories for mycorrhiza, they have their greatest importance for trees (and occur most frequently) at low or intermediate levels of plant nutrients. Mycorrhizal fungi can be expected to decrease, as has also been observed (Wästerlund 1982; Jansen and van Dobben 1987). There will be some leaching of nitrate, but in this phase probably less than what is added from outside. On sites low in some other nutrient, deficiency symptoms may appear, mainly due to lack of boron, phosphorus, or magnesium, when biomass increases.

The next stage in the development will be "nitrogen saturation" as defined in Sect. 4.1. This stage can be attained in two ways: (1) by serious disturbance of the stand (e.g., clear-felling) and may then be at least partly reversible, if it is possible to establish a new forest stand, and (2) by exceeding the threshold where growth is not further increased in the existing vegetation and accumulation in the soil fails to keep pace with the nitrogen input from outside.

A mathematical model constructed by Ågren (1983) allows us to calculate the time required to attain stationary needle biomass, i.e., when tree growth is no longer increased with increased nitrogen deposition (Fig. 5.7). Despite the simplifications in the model, the results agree reasonably well with the data available from the Swedish optimum nutrition experiments discussed in Sect. 4.4. The experimental results suggest that plots receiving 30 kg N ha^{-1} $year^{-1}$ (in addition to a low atmospheric deposition) over the last decade, with higher additions in the first years, are still in the first phase (not yet saturated). Plots having received twice as much nitrogen show signs of approaching saturation (soil acidification and elevated nitrate concentrations in the mineral soil), while plots with 90 kg annually show a certain decline in growth. However, as the various nitrogen regimes affect not only tree nutrition but also soil acidity (as does anthropogenic nitrogen deposition), the detailed interpretation of the experimental results is complicated. According to Ågren and Bosatta (1988), the soil, which in their model is the forest floor, is nitrogen-saturated earlier than the stand, a result of the modelling which appears confirmed in the experiment. Soil losses of nitrate with percolating water seem to occur before any serious decline symptoms in the stand. In addition, at higher nitrogen regimes only a smaller part of the added nitrogen could be recovered in soil and stand, according to preliminary estimates (Sect. 4.4; Fig. 4.14).

The consequences of clear-felling or other disturbances on sites already nitrogen-saturated can be expected to be even less desirable than in ecosystems still in the nitrogen-accumulating phase, where we foresaw some reversibility. The vegetation on the clear-felling will consist of nitrophilous plants, which may hold back some nitrogen, but presumably not more than a fraction of what is mineralized. Microbial immobilization of nitrogen, which is an important process in ecosystems normally nitrogen-limited (Vitousek and Matson 1984), would be expected to decrease in ecosystems approaching saturation. Soil acidification (due to both sulphur and nitrogen deposition) does not favour the normally most common nitrifying bacteria, but we have earlier stated that nitrification can also occur under acid conditions. Moreover, soil pH usually increases temporarily in a mor layer upon clear-felling, presumably because acid organic matter is decomposed (Nykvist and Rosén 1985). On the other hand, excessive nitrification would further acidify the soil and

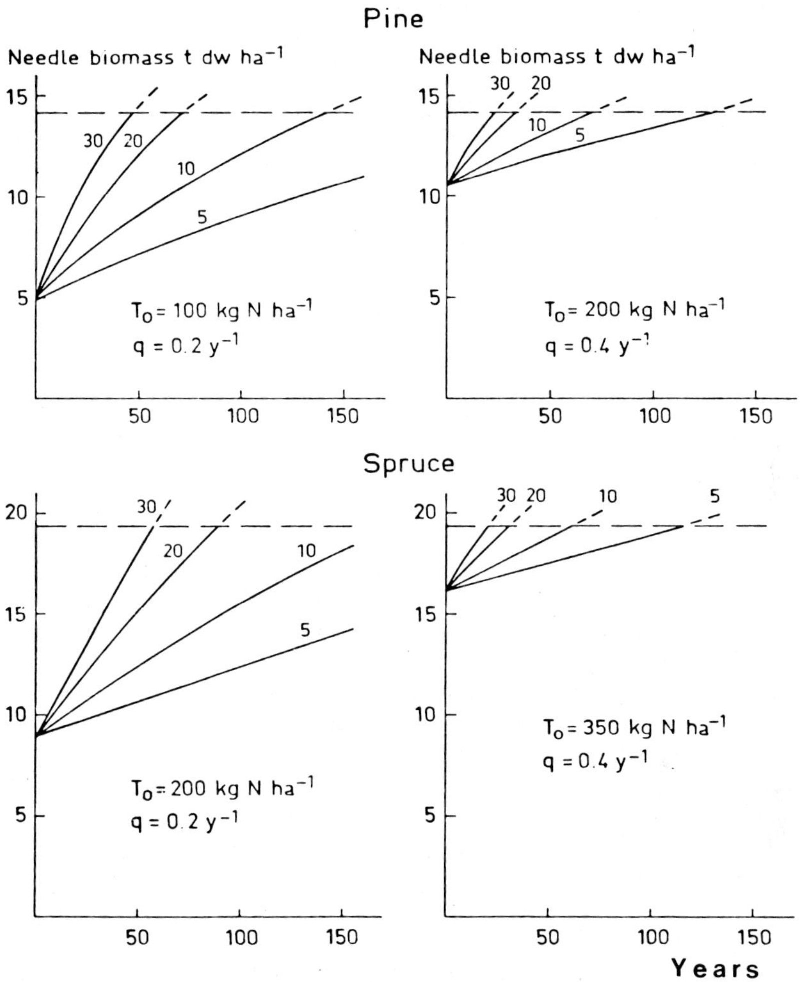

Fig. 5.7. The development of the stationary needle biomass with different deposition rates of nitrogen (5, 10, 20, 30 kg N ha^{-1} year^{-1}) starting from poor and medium pine and spruce stands, respectively. The initial pool of rapidly turning-over nitrogen (in foliage and forest floor) is designated T_0. The rate of mineralization (q) is assumed to be half as high on poor sites (*left*) as on medium sites (*right*) (Ågren 1983)

mobilize aluminium and other metals deposited in the accumulation horizon.

We expect that nitrogen-limited forests on poor sites initially react slower on increased nitrogen deposition, if compared with richer forest types with mull humus and a nitrification potential already under natural conditions. This is also the case in the Ågren model (Fig. 5.7). There are at least two main reasons: (1) plant species adapted to severe nitrogen limitation do not change their growth as much as more demanding species (Fig. 5.1; Chapin et al. 1986), and (2) soil immobilization of nitrogen may be very strong in nitrogen-limited ecosystems (Chapin et al. 1986).

However, when the originally poorer forests reach the saturation stage, or even before that, they may change into less stable systems, where disturbances may result in even more serious imbalances than in types richer from the beginning.

The same general principles would apply to heathlands as well as grasslands, the former being more similar to poor forests with mor and the latter to richer forests with mull.

A certain parallel with the possibly higher sensitivity of poorer types to eutrophication can be found on the aquatic side, where the reaction to sewage may lead to more unbalanced conditions in oligotrophic lakes than in more eutrophic ones.

5.3 Acute Damage to Plants Associated with Emissions of Nitrogen Compounds

As the main scope of this volume is to discuss the ecological role of nitrogen as a plant nutrient and to describe possible consequences of increased nitrogen supply mediated by the soil, direct damage by gaseous compounds containing nitrogen or otherwise nitrogen related compounds (ozone) will only be briefly mentioned, with some references and a few comments. For more detailed information the reader is referred to the extensive literature in the field (Mathy 1988; Unsworth and Ormrod 1982; WHO 1987). The proceedings from the ECE *Critical Levels Workshop* in Bad Harzburg, FRG in March 1988 are presently available as a Draft Report (Anonymous 1988).

5.3.1 Damage Caused by NO_x or HNO_3

The burning of fuels or simply heating of air inevitable leads to the formation of nitrogen oxides, of which NO is the most common one in emissions. However, NO is rapidly oxidized to NO_2 in air, which can be further oxidized to nitric acid, HNO_3. Both compounds are reactive and therefore have a more limited residence time in air than, for instance, sulphur dioxide. Yet, like sulphuric acid formed from sulphur dioxide, nitric acid may also be dissolved in aerosol droplets and thereby have a longer residence time in the atmosphere and longer travel distances from source areas. In contact with vegetation, both NO_2 and gaseous or dissolved HNO_3 are eagerly absorbed and cause damage if the concentrations are high enough (WHO 1987). The presence of sulphur dioxide lowers the threshold for damage.

5.3.2 Damage Caused by Photochemical Oxidants, in Particular Ozone

In the lower atmosphere (the troposphere), ozone is formed in strong sunlight if nitrogen oxides (NO_x) and organic substances such as terpenes or products from incomplete combustion are present simultaneously.

Ozone damage has been a popular hypothesis to explain the extensive forest damage in Central Europe, starting, or at least first noted, in the beginning of the 1980s (for references, see Roberts et al. 1983; Guderian et al. 1985; Krause et al. 1985; Anonymous 1987). It is also true that episodes with ozone concentrations at or close to experimentally established threshold values for vegetation damage also occur outside the most polluted regions, e.g., in Scandinavia. Intensive work is going on in this field, but as ozone damage is related to nitrogen in a rather indirect way, the reader is referred to the extensive literature on ozone formation in the lower atmosphere and ozone damage in plants.

Next to ozone, peroxyacetyl nitrate (PAN) is the best-known photochemical oxidant. It is formed under similar conditions to ozone, usually in lower concentrations. It is reported that trees are less sensitive than herbs to PAN (Guderian et al. 1985).

5.3.3 Damage Associated with Gaseous Ammonia

It has been known for some time (Lemon and van Houtte 1980, with further references) that plants can take up gaseous ammonia from the air, preferentially by stomata (and also that cells, in particular senescent ones, may lose ammonia the same way, as mentioned in Sect. 2.2.1). With increasing emission of acid substances into the atmosphere, it seems to have been taken for granted that there is little free ammonia in the air in polluted areas. However, a main source of ammonia is intensive agriculture with large livestock. High air concentrations of ammonia have been measured by Nihlgård (1985) in forests adjacent to such farms, with the concentration gradient indicating that the forest acts as a sink for ammonia. The extremely high nitrogen deposition in the Netherlands also contains a large proportion of ammonium nitrate, considered to originate in animal husbandry and other agricultural activities (Erisman et al. 1987). In the cases studied by Nihlgård, spruce needles accumulated nitrogen well above optimum levels (cf. Sect. 4.4) and, in addition, became overgrown with green algae, not a normal feature in south-Swedish forests, and presumably not good for needle function. It is well known that free ammonia as well as high concentrations of ammonium are poisonous to plant cells, but so far it is not fully proven that the tree damages observed are due to direct poisoning; increased sensitivity to other stresses at above-optimum nitrogen levels is an alternative explanation.

5.3.4 Damage to Roots Associated with Nitrification-Caused "Acid Surges" in Soil

As this phenomenon belongs to indirect rather than direct damage, it has been discussed earlier (Sect. 2.2.2) and is listed here only in order to make the list of pathological influences associated with nitrogen compounds more complete.

5.3.5 Damage by Climatic Stress Aggravated by Increased Sensitivity in Nitrogen-Rich Plants

This possibility has already been mentioned in discussing the unbalanced nutrition caused by high nitrogen inputs (Sect. 4.4). In specific cases it may be difficult to determine the extent of increased sensitivity to climatic stresses due to a high nitrogen supply and the importance of a low supply of potassium, boron, or some other element. Yet it should be remembered that the evolution of our forest trees and accompanying plant species has taken place in an environment where nitrogen almost always has been a limiting factor. Ample supply of nitrogen is not a normal condition for these species, and increased nitrogen supply may lead to morphological (e.g., root/shoot ratio), anatomical (e.g., cell size), and chemical changes. It is thus easy to understand that access to more nitrogen may aggravate stress induced by some climatic factor such as frost or drought.

5.3.6 Damage by Grazing Animals or Pests, Associated with Changes in Palatability or Chemical Defense

This possibility is much discussed, but hard facts are scarce. As insect bodies generally contain higher concentrations of nitrogen than the plant tissues they eat, it is easy to believe that the amount of nitrogen ingested is a critical factor for their development. Alternatively, the critical factor may be one or more essential amino acids (White 1984). However, there are cases described both where pests preferentially attack well-nourished plants and where nutrient-deficient plants are attacked. Many observed relations concern rather indirect relationships, where the important thing may be growth rate, cell size, or pressure in resin ducts rather than chemical defense substances or food quality measured, e.g., such as protein or amino acid content.

There seems to be a fairly general agreement that increased nitrogen level in phloem sap increases its nutritional value for aphids and other sucking insects (Raven 1983). These animals apparently have an excess of soluble carbohydrates, when excreting honeydew rich in sugar. Similarly, a lower C/N ratio (= higher protein concentration) in foliage is likely to improve the nutritional value for leaf-eating caterpillars, though most of the evidence for this hypothesis is indirect.

On the other hand, plant defense mechanisms may function in a way which overrules the simple direct effects. If increased nitrogen supply increases plant growth efficiency, which only happens at suboptimum nitrogen supply, trees may survive insect attacks better with than without fertilization (Waring and Pitman 1983). Chemical defense mechanisms are often classified in two groups: nitrogen-based and carbon-based. It is easy to understand that concentrations of nitrogen-based compounds may increase with increasing nitrogen supply. Carbon-based defense substances such as tannins and other polyphenolic compounds are more likely to decrease with increasing nitrogen level, but there are also other controlling mechanisms, i.a., genetic factors. As carbon-based defense mechanisms are more common in trees and also in plants adapted to sites of low fertility, nitrogen enrichment and a possible nitrogen saturation pose a potential threat, viz., that these plant species will become more susceptible to insect damages, and possibly also to other pests.

For further information, the reader is referred to the extensive literature on pathology (e.g., Mattson 1980; Mattson and Scriber 1987; Waring and Pitman 1983; White 1984; Gulmon and Mooney 1986; Brodbeck and Strong 1987; Hain 1987).

Most work on the relations between forest tree nutrition and pest attacks have dealt with insect pests. It is therefore interesting that Lambert (1986) found that infection of *Pinus radiata* needles by the fungus *Dothistroma* increased with nitrogen fertilization. The infection rating was linearly related to the foliage arginine concentration.

5.4 Interactions Between Effects of Anthropogenic Nitrogen Emissions and Other Human Impacts on Terrestrial Ecosystems

5.4.1 Changes in Land Use

We have already mentioned that some of the recent changes in land use in Europe (and also in other parts of the world) interfere with the nitrogen regime on a site. The succession on abandoned field, pastures, or hay-meadows implies what we have called a self-eutrophication, as ecosystems developing towards some sort of climax vegetation usually have fewer outputs of nutrients than the managed system and still get inputs from weathering, deposition, nitrogen fixation, and, sometimes, flushing water. Many ecologists consider the increasing conservation of nutrients during a succession as a kind of law of Nature. In any case, this development is widespread and would, under undisturbed conditions, lead to some sort of steady state. This "climax" is not necessarily a constant end stage, but might well oscillate over time, as in a savanna burning annually or a boreal forest burning once every century. There is little doubt that nitrogen is one of the plant nutrients whose accumulation over the succession is most important, a fact utilized by "primitive" farmers in various parts of the world in the practice of shifting cultivation. It is not known what the end product of succession in land without human interference (cropping, grazing, forest harvesting) will look like in areas with high nitrogen deposition. We have discussed changes in tree layer and other vegetation, as well as in soils in previous sections, but it is difficult to draw safe conclusions, particularly concerning the degree of stability/instability of the new systems created. Such a discussion requires information on the interactions dealt with in the next two sections, information which at present is far from complete.

5.4.2 Management Impacts on Forest Ecosystems Enriched with Nitrogen

Most forests in those parts of the world affected by nitrogen emissions are managed. This means, on the one hand, that species composition is human-controlled, often to monocultures. On the other hand, it means regular harvest of timber and possibly other parts of the biomass, which, i.a., results in a removal of nitrogen, as well as of other nutrients. Further impacts may be more or less intensive site preparation before planting and/or fertilization at establishment or later.

Introduction of coniferous monocultures has strong effects on several ecosystem functions. The litter is usually more acid than that in the replaced deciduous woodlands or abandoned farmland. Conifer litter is also rich in lignin and other polyphenolic compounds, which are unpalatable for many soil animals and which retard decomposition by microorganisms (Melillo et al. 1982; Berg 1986). All this favours a decomposition of the mor type and podzolization. In Sect. 5.2 we have already discussed the effect of nitrogen additions on mull and mor systems.

According to commonly accepted pedological principles, soil profiles, with their static and dynamic characteristics, should be considered in terms of their functional relations with the ecosystems. The theory of soil-forming factors (Jenny 1941, 1980) is an expression of this approach. A natural but not easily proven conclusion would be that soils in transition, i.e., not in equilibrium with the soil-forming factors, would be more sensitive to further disturbance such as acid or nitrogen deposition than soils typical for the natural forest types of the region.

Podzols with mor humus are typical for coniferous forests of the boreal region, and cambisols and luvisols for deciduous forests of the temperate region, and there are reasons to believe that long established zonal ecosystems are less sensitive, at least to acid deposition (Malmer 1974), than ecosystems in transition, due to planting of conifers outside their natural range. Whether this is true as well for effects of nitrogen deposition is not proven, but it is clear that mor type humus from typical boreal sites shows a considerable resistance to nitrification, even where the plots have received moderate annual applications of ammonium nitrate (Table 4.2, Fig. 4.14; Tamm 1974; Popovic 1977; Nilsson et al. 1988).

Concerning the temperate hardwood region, Ulrich (1983) reports from Solling, a formerly beech-dominated forest area in Germany, now partly under spruce canopy, that what he calls discoupling between nitrate formation and root uptake is particularly evident beneath the introduced spruce. In Ulrich's case much of the nitrification appears to take place in mineral soil horizons, and we have already regretted the fact that most laboratory and field studies on nitrification in forest soils only concern the humus layer and possibly the uppermost mineral soil.

With the use of monocultures follows the need for clear-felling and often for site preparations. These operations usually lead to losses of nitrogen from the site by leaching and denitrification. While such changes between accumulating and dissipating phases also occur in natural ecosystems (Sect. 3, see also Bormann and Likens 1979), the shifts are often much more pronounced in intensively managed forests, due to shorter rotations, faster growth, occasional fertilization, and soil cultivation in the regeneration phase.

In the long run (over rotations) a balance would be expected between nitrogen inputs and outputs. Biomass removal may compensate for part of the atmospheric deposition, but as the tree stems are comparatively low in nitrogen content the removal is of moderate importance except in cases of whole-tree utilization, or where the slash is wind-rowed and burnt. Removal due to harvesting also removes plant nutrients other than nitrogen and can thus easily lead to unbalanced nutrition.

As soil disturbances favour organic matter decomposition and nitrogen mineralization (Sect. 3.3), nitrogen losses by leaching from mechanically disturbed sites can be important. Our general conclusion is that the growth and yield of managed forests are not immediately threatened by the present increases in nitrogen deposition except in areas with extreme conditions, such as parts of the Netherlands (Boxman et al. 1988). We then make the somewhat uncertain assumption that direct damages such as those discussed in Sect. 5.3 are of less importance than the soil-mediated effects. Yet there is a great risk that man-made forests with a good supply of nitrogen will be less stable in various ecological respects. This implies potential risks both on the site in question and for other ecosystems, in particular those receiving the runoff from the forest.

5.4.3 Interactions Between Deposition of Nitrogen and Other Air Pollutants

Increased nitrogen deposition is only a part of the air pollution problems. Until now, sulphur compounds (sulphur dioxide, sulphuric acid, and acid sulphates) have been emitted in larger quantities than acidifying nitrogen compounds. The better possibilities of controlling emissions of sulphur than of nitrogen may change the present ratio, but most of the soil and water acidification which has occurred up until now can be ascribed to sulphur rather than to nitrogen, partly because most forest ecosystems in sensitive areas have been nitrogen-limited and thus able to absorb and accumulate deposited nitrogen in reduced form up to the present time.

We have already discussed relations between soil nitrogen turnover and soil acidity (Sect. 5.2). Increased soil acidity generally favours the mor type of decomposition, as described earlier. This would temporarily decrease nitrification, but not necessarily when nitrogen saturation is attained. High concentrations of absorbed and dissolved ammonium ions, which would result at nitrogen saturation in the case of impeded nitrification, are definitely undesirable, as experiences from the Netherlands have already shown (Boxman et al. 1988).

Two consequences of increased soil acidity combined with a rich supply of nitrogen appear obvious: (1) Rooting becomes shallower when the mineral soil is acidified and most of the nutrient release takes place in the forest floor instead of in the mineral soil-mixed mull, and ions in soil water in the mineral horizons will be more easily leached when biological retention by root uptake decreases. The mineral soil will also be richer in free aluminium and iron with increasing acidity, which is clearly unfavourable for roots of a number of species at least and, in addition, may lead to lower phosphorus availability due to complexing with aluminium and iron. Shallow rooting decreases drought resistance, as only roots in deeper horizons can take up water under prolonged drought periods. (2) The occurrence of nitrate ions together with sulphate ions in the soil solution further increases the leaching of cations, as anions are always accompanied by cations in the mass-flow down in the soil.

The probable consequences of soil acidification combined with increased nitrogen deposition will most probably be unbalanced nutrition over large areas, as has already been reported in the case of magnesium in southern Germany (Zech and Popp 1983; Hüttl 1985). Both increased leaching of magnesium from needles (Johnsen et al. 1987) and decreases in soil stores of exchangeable magnesium (Tamm and Popovic 1989) have been observed as a consequence of experimental acid deposition.

Critical loads of deposition of both sulphur and nitrogen have been discussed in considerable detail recently (Nilsson and Grennfelt 1988). There is still some discussion on the definition of the concept, particularly on how much change in the environment can be allowed, and whether the critical value has to be set with respect to the most sensitive ecosystem or ecosystem component. However, there seems to be agreement that the acidification load is determined equally by sulphur and nitrogen deposition, as soon as nitrogen saturation is attained. A number of methods of estimating critical loads for sulphur (or rather for acid deposition) agree reasonably well (Nilsson and Grennfelt 1988). Critical loads for nitrogen as an eutrophication factor have not been studied very long, but there is a fairly good agreement between the theoretical approach by Ågren and Bosatta (1988) and the experimental results from the present author's group, partly reviewed in Sect. 4.4. Vegetation studies from the Netherlands and Germany (Sect. 5.1) indicate that the critical load with respect to sensitive plant species is already exceeded over whole regions.

Other consequences of air pollution are connected with deposition of heavy metals, and of the higher solubility and hence toxicity of these metals in acidified soils. It is well known that high concentrations of heavy metals interfere with litter decomposition and soil biological processes in general (Rühling and

Tyler 1973; Freedman and Hutchinson 1980). It is not clear what the result of all these interactions will be under different conditions, e.g., to what extent presence of nitrate will increase plant uptake of heavy metals and thereby increase heavy metal cycling, but several possibilities exist. Increased outflow of heavy metals (and aluminium) from soils is to be expected from the combination of high acidity and occurrence of nitrate ions in the soil solution.

Since heavy metal pollution, especially in combination with soil and water acidification, constitutes one of our most threatening environmental problems for the future, it deserves a much more extensive discussion than is possible in this context.

6 Conclusions

One main purpose of this book is to stimulate discussion on the serious problems connected with one important aspect of the recent changes in chemical climate, viz., the increased atmospheric deposition of nitrogen compounds. An early mimeographed version served as a background paper at the Nordic/UN-ECE Workshop: *Critical Loads for Sulphur and Nitrogen*, March 21–24, Skokloster, Sweden. The Report from this workshop is now available (Nilsson and Grennfelt 1988) and several of the papers in that volume are quoted here.

The conclusions drawn at the Skokloster Workshop were agreed upon by a number of experts from many countries and, therefore, the discussion here can be limited to a few more personal conclusions and statements, which hopefully will make it clear that continued studies within a broad range of site types are necessary, if we are to be able to fully understand the ongoing changes in the nitrogen status of ecosystems affected by emissions of nitrogen compounds as a result of human activities. An understanding of the changes on an ecosystem level is essential for our possibilities of finding the most efficient and economic ways to counteract the changes or at least alleviate the adverse effects. Yet some of the consequences of nitrogen enrichment in terrestrial ecosystems are already so obvious that action should be taken immediately: tree damage near heavy traffic or large animal farms, groundwater nitrate in intensively farmed areas, with subsequent outflow of nitrogen with rivers and eutrophication of coastal waters.

Some points to consider are:

1. The present enrichment of nitrogen by human activities, both agricultural and industrial, is considerable in large regions of the earth. Even if, in some countries, legislation has introduced restrictions on large industries, car exhausts, and fertilizer use, these restrictions are clearly insufficient for a strong and rapid decrease of present emissions from combustion and agricultural use of nitrogen.
2. Nitrogen enrichment at the present scale is an unprecedented event in geologic and ecologic history. Enriched sites do occur naturally, but seldom cover large areas, and depend on "donor ecosystems" for their persistence. Thus we lack experience from the past to judge the ecological consequences of extensive nitrogen enrichment.
3. A number of the ecological consequences have so far been positive for plant production on both cultivated land and in managed forests: increased yield or reduced need for nitrogen fertilizers.

4. Adverse ecological effects of supra-optimum nitrogen levels are now being recorded in areas with high deposition. In addition, nitrogen deposition, especially when the accumulation exceeds the limits of tolerance of the ecosystem, will contribute considerably to soil acidification and presumably to increased forest damage.
5. Many plant and animal species adapted to an environment where nitrogen is a limiting factor will not survive under the new conditions created by anthropogenic nitrogen enrichment. We are already witnessing the beginning of an extinction process, which concerns both species and whole ecosystems, such as heathlands, some wetlands, and some forests. Even if small patches of lands can be managed to remove nitrogen by harvesting, there is no guarantee that we can preserve ecosystems and species in this way, not knowing all the factors important for endangered species. On the contrary, ecological experience tells us that it is impossible to preserve ecological diversity in small and often isolated patches, "islands" surrounded by an environment hostile to the species we wish to preserve.
6. It is important to concentrate further research on functions of ecosystems with high loads of nitrogen, and to make this research truly ecosystem-directed, i.e., interdisciplinary and integrated. A variety of approaches are needed: comparative studies along pollution gradients, short-term and long-term experiments on the effects of single factors and their interactions, and retrospective studies using both "biological archives" and earlier records and investigations. No single method can solve the complex problems of ecosystem reactions even to a single factor, much less the problems arising from the various sorts of chemicals emitted as a result of modern human activities.

References

Aber JD, Mellillo JM (1982) Fortnite: a computer model of organic matter and nitrogen dynamics in forest ecosystems. Agric Res Bull R3130, Univ Wisc, Madison

Abrahamsen G (1980) Acid precipitation, plant nutrients and forest growth. In: Drablös D, Tollan A (eds) Ecological impact of acid precipitation. Proceedings. Oslo-Ås, pp 58–63

Ågren GI (1983) Model analysis of some consequences of acid precipitation on forest growth. In: Ecological effects of acid precipitation. SNV PM Rep, Stockholm, 1636:233–244

Ågren GI (1986) (ed) Predicting consequences of intensive forest harvesting on long-term productivity. Swed Univ Agric Sci Dept Ecol Environ Res Rep 26, Uppsala, 205 pp

Ågren G, Bosatta E (1988) Nitrogen saturation of the terrestrial ecosystems. Environ Pollut 54:185–197

Albrektson A, Aronsson A, Tamm CO (1977) The effect of fertilization on primary production and nutrient cycling in the forest ecosystem. Silva Fenn 11:233–239

Alexander IJ (1983) The significance of ectomycorrhizas in the nitrogen cycle. In: Lee JA, McNeill S, Rorison IH (eds) Nitrogen as an ecological factor. 22nd Symp Brit Ecol Soc, Oxford, pp 69–94

Al-Mufti MM, Sydes CL, Furness SB, Grime JP, Band SR (1977) A quantitative analysis of shoot phenology and dominance in herbaceous vegetation. J Ecol 65:759–791

Anonymous (1983) The nitrogen cycle of the United Kingdom. The Royal Society of London, London, 264 pp

Anonymous (1987) Direct effects of dry and wet deposition on forest ecosystems. Proceedings from a workshop October 19–23, 1986, Lökeberga, Sweden. CEC Air Pollution Report No 4, EUR 11264, 300 pp

Anonymous (1988) ECE Critical Levels Workshop. Final Draft Report. Bad Harzburg, FRG, 146 pp

Aronsson A (1980) Frost hardiness in Scots pine. II. Hardiness during winter and spring in young trees of different mineral status. Stud For Suec 155:1–27

Aronsson A (1983) Growth disturbances caused by boron deficiency in some fertilized pine and spruce stands on mineral soil. Commun Inst For Fenn 116:116–122

Aronsson A (1985a) Changes in mineral content of forest trees after fertilization. Skogsfakta Suppl 5:55–60 (in Swedish)

Aronsson A (1985b) Indications of stress at unbalanced nutrient contents of spruce and pine. K Skogs- o Lantbruks akad Tidskr Suppl 17:40–51 (Swedish with English summary)

Aune EI, Dahl E, Løes AK (1989) A comparison of conifer forests and their soils in relation to acid precipitation in Central Norway, South Norway and Schwarzwald. Medd norske Skogforsøksvesen 42(1):133–146

Axelsson B (1985) Biomass dynamics in the nutrition experiment at Stråsan. K Skogs- o Lantbruksakad Tidskr Suppl 17:30–39 (Swedish with English summary)

Barkman JJ (1969) The influence of air pollution on bryophytes and lichens. In: Air pollution. Proceedings of the First European Congress on the influence of air pollution on plants and animals. Wageningen 1968, pp 197–209

Berdén M, Nilsson SI, Rosén K, Tyler G (1987) Soil acidification – extent, causes and consequences. SNV Report 3292, Solna, Sweden, 161 pp

Berg B (1986) Nutrient release from litter and humus in coniferous forest soils – a minireview. Scand J For Res 1:359–370

Berg B, Staaf H (1981) Leaching, accumulation and release of nitrogen in decomposing forest litter. Ecol Bull (Stockholm) 33:163–178

Bergqvist B (1987) Leaching of metals from forest soils as influenced by tree species and management. For Ecol Manage 22:29–56

Bjarnason S (1987) Nitrogen dynamics in the long-term soil fertility experiments in Malmöhus County (Swedish with English summary). K Skogs- o Lantbruksakad Tidskr Suppl 19:65–70

Björkman E (1943) Über die Bedingungen der Mykorrhizabildung bei Kiefer und Fichte. Symb Bot Ups 6(2):190 pp

Björkman E (1960) *Monotropa hypopitys* L. – an epiparasite on tree roots. Physiol Plant 13:308–327

Björkman E (1970) Forest tree mycorrhiza – the conditions for its formation and the significance for tree growth and afforestation. Plant Soil 32:589–610

Bormann FH, Likens GE (1967) Nutrient cycling. Science 155:424–429

Bormann FH, Likens GE (1979) Pattern and process in a forested ecosystem. Springer, Berlin Heidelberg New York, 253 pp

Bowen GD, Smith SE (1981) The effects of mycorrhiza on nutrient uptake by plants. In: Clark FE, Rosswall T (eds) Terrestrial nitrogen cycles. Ecol Bull Stockholm 33:237–247

Boxman AW, Sinke RJ, Roelofs JGM (1986) Effects of NH_4^+ on the growth and K^+ (^{86}Rb) uptake of various ectomycorrhizal fungi in pure culture. Water Air Soil Pollut 31:517–522

Boxman D, van Dijk H, Roelofs J (1988) Critical loads for nitrogen with special emphasis on ammonium. In: Nilsson J, Grennfelt P (eds) Critical loads for sulphur and nitrogen. Miljörapport 1988, Copenhagen, 15:295–322

Braekke FH (1977) Fertilization for balanced mineral nutrition of forests on nutrient-poor peatland. Suo 28:53–61

Brantseg A (1966) Forest fertilization and game animals. Jakt, Fiske, Friluftsliv 95:216–219 (in Norwegian)

Breznak JA, Brill WJ, Mertins JW, Coppel HC (1973) Nitrogen fixation in termites. Nature (Lond) 244:577–579

Brimblecombe P, Stedman DH (1982) Historical evidence for a dramatic increase in the nitrate component of acid rain. Nature (Lond) 298:460–462

Brodbeck B, Strong D (1987) Amino acid nutrition of herbivorous insects and stress to host plants. In: Barbosa P, Schultz JC (eds) Insect outbreaks. Academic Press, San Diego, pp 347–364

Bryant JP, Chapin FS III, Klein DR (1983) Carbon/nutrient balance of boreal plants in relation to vertebrate herbivory. Oikos 40:357–368

Burgtorf H (1981) The effect of supply of plant nutrients to some pine stands on poor and cold sites (Swedish with summary in English). Rep For Ecol For Soils 37:1–34, SLU, Uppsala

Carmean WH (1967) Soil refinements for predicting black oak site quality in southeastern Ohio. Proc Soil Sci Soc Am 31:805–810

Chapin FS III (1980) The mineral nutrition of wild plants. Annu Rev Ecol Syst 11:233–260

Chapin FS III, Kedrowski RA (1983) Seasonal changes in nitrogen and phosphorus fractions and autumn retranslocation in evergreen and deciduous taiga trees. Ecology 64:374–391

Chapin FS III, Vitousek PM, Van Cleve K (1986) The nature of nutrient limitation in plant communities. Am Nat 127:47–58

Cody ML, Mooney HA (1978) Convergence versus nonconvergence in Mediterranean-climate ecosystems. Annu Rev Ecol Syst 9:265–321

Coe M (1983) Large herbivores and food quality. In: Lee JA, McNeill S, Rorison IH (eds) Nitrogen as an ecological factor. 22nd Symp Br Ecol Soc Oxford, pp 345–368

Cole DW, Gessel SP, Dice SF (1967) Distribution and cycling of nitrogen phosphorus, potassium and calcium in a second-growth Douglas Fir ecosystem. In: Young HE (ed) Symposium on primary productivity and mineral cycling in natural ecosystems. Univ Maine Press, Orono, pp 197–232

Cole DW, Rapp M (1981) Elemental cycling in forest ecosystems. In: Reichle DE (ed) Dynamic properties of forest ecosystems. Cambridge University Press, Cambridge, pp 341–409

Collins NM (1983) The utilization of nitrogen resources by termites (Isoptera). In: Lee JA, McNeill S, Rorison IH (eds) Nitrogen as an ecological factor. 22nd Symp Br Ecol Soc Oxford, pp 345–368

Crocker RL, Major J (1955) Soil development in relation to vegetation and surface age at Glacier Bay, Alaska. J Ecol 43:427–448

Cromack K, Sollins P, Todd RL, Fogel R, Todd AW, Fender WM, Crossley ME, Crossley DA (1977) The role of oxalic acid and bicarbonate in calcium cycling by fungi and bacteria: some possible implications for soil animals. Ecol Bull Stockholm 25:246–252

Cromack K, Sollins P, Graustein WC, Seidel K, Todd AW, Spycher G, Ching YL, Todd RL (1979) Calcium oxalate accumulation and soil weathering in mats of the hypogenous fungus *Hysterangeum crassum*. Soil Biol Biochem 11:465–468

Dahl E, Gjems O, Kielland-Lund J (1967) On the vegetation types of Norwegian conifer forests in relation to the chemical properties of the humus layer. Medd norske Skogforsøksvesen 23:503–531

Darwin Ch (1881) On the formation of vegetable mould through the actions of worms. (Reprinted 1946, Faber and Faber, London)

Davidson EE, Swank WT (1986) Environmental parameters regulating gaseous nitrogen losses from two forested ecosystems via nitrification and denitrification. Appl Environ Microbiol 52:1287–1292

Dirkse GM, van Dobben HF (1989) Effects of experimental fertilization on forest undergrowth in young stands of Scots pine in Sweden. In: Sjögren E (ed) Forests of the world. Diversity and dynamics (Abstracts). Stud Plant Ecol 18:62–64

Dyck WJ, Messina MG, Hunter IR (1986) Predicting the nutritional consequences of intensive harvesting on site productivity. In: Ågren GI (ed) Predicting consequences of intensive harvesting on long-term productivity. Swed Univ Agric Sci Dept Ecol Environ Res Rep Uppsala 26:9–29

Ellenberg H (1977) Stickstoff als Standortsfaktor, insbesondere für mitteleuropäische Pflanzengesellschaften. Oecol Plant 12:1–22

Ellenberg H (1979) Zeigerwerte der Gefässpflanzen Mitteleuropas. Scr Geobot 9:122

Ellenberg H Jr (1985) Veränderungen der Flora Mitteleuropas unter dem Einfluss von Düngung und Immissionen. Schweiz Z Forstwes 136:19–39

Ellenberg H Jr (1988) Eutrophierung – Veränderungen der Waldvegetation – Folgen für den Reh-Wildverbiss und dessen Rückwirkungen auf die Vegetation. Schweiz Z Forstwes 139:261–282

Ellenberg H Jr (in press) Eutrophication as a significant background problem for European wildlife. In: Afkew M (ed) EEC Symposium on oil rape seed and wildlife. Brussels, Sept 1988

Emanuelsson U (1989) The relationship of different agricultural systems to the forest and woodlands of Europe. In: Salbitano F (ed) Human influence on forest ecosystems development in Europe. Pitagora Editrice, Bologna, pp 169–178

Ericsson A (1978) Seasonal changes in translocation of ^{14}C from different age-classes of needles on 20-year-old Scots pine trees (*Pinus sylvestris*). Physiol Plant 43:351–358

Ericsson A (1979) Effects of fertilization and irrigation on the seasonal changes of carbohydrate reserves in different age-classes of needles on 20-year-old Scots pine trees (*Pinus sylvestris*). Physiol Plant 44:270–280

Erisman JW, Vermetten AW, Pinksterboer EF, Asman WAH, Waijers-Ypelaan A, Slanina J (1987) Atmospheric ammonia-distribution equilibrium with aerosols and conversion rate to ammonium. In: Asman WAH, Diederen HSMA (eds) Ammonia and acidification. Eurosap Symposium Proceedings, Bilthoven, The Netherlands, pp 59–72

Falkengren-Grerup U (1986) Soil acidification and vegetation changes in deciduous forest in southern Sweden. Oecologia 70:339–347

Falkengren-Grerup U, Eriksson H (in press) Changes in soil, vegetation and forest yield between 1947 and 1988 in beech and oak sites of southern Sweden. For Ecol Manage

Farquhar GD, Firkl PM, Wetselaar R, Weir B (1980) On the gaseous exchange of ammonia between leaves and the environment: determination of the compensation point. Plant Physiol 66:710–714

Farquhar GD, Wetselaar R, Weir B (1983) Gaseous nitrogen losses from plants. In: Freeney JR, Simpson JR (eds) Gaseous loss of nitrogen from plant-soil systems. Dev Plant Soil Sci 9:159–180

Field CB, Mooney HA (1986) The photosynthesis-nitrogen relationship in wild plants. In: Givnish T (ed) On the economy of plant form and function. Cambridge University Press, pp 22–55

Flaig W, Beutelspacher H, Rietz E (1975) Chemical composition and physical properties of humic substances. In: Gieseking JE (ed) Soil components. I. Organic compounds, p 1–212. Springer, Berlin Heidelberg New York

Focht DD, Verstraete W (1977) Biochemical ecology of nitrification and denitrification. Adv Microb Ecol 1:135–214

Fogelfors H, Steen E (1982) Vegetation changes during 25 years in landscape conservation experiments near Uppsala, Sweden. SNV PM 1623, 66 pp (in Swedish)

Freedman B, Hutchinson TC (1980) Smelter pollution near Sudbury, Ontario, Canada, and effects on forest litter decomposition. In: Hutchinson TC, Havas M (eds) Effects of acid precipitation on terrestrial ecosystems. NATO Conference series 1, Ecology Plenum, New York, pp 395–434

Gadgil P, Gadgil R (1975) Suppressed litter decomposition by mycorrhizal roots of *Pinus radiata*. NZJ For Sci 5:33–41

Gärdenfors U, Westermark T, Emanuelsson U, Mutvei H, Waldén H (1988) Use of land-snail shells as environmental archives: preliminary results. Ambio 17:347–349

Gebauer G, Rehder H, Wollenweber B (1988) Nitrate, nitrate reduction and organic nitrogen in plants from different ecological and taxonomic groups of Central Europe. Oecologia 75:371–385

Gerhardt K, Kellner O (1986) Effects of nitrogen fertilizers on the field- and bottomlayer species in some Swedish coniferous forests. Medd Växtbiol Inst Uppsala Univ 1986:1, 47 pp

Gorham E (1953) The development of the humus layer in some woodlands of the English Lake District. J Ecol 41:123–152

Gorham E, Vitousek PM, Reiners WA (1979) The regulation of element budgets over the course of terrestrial ecosystem successions. Annu Rev Ecol Syst 10:53–84

Grennfelt P, Hultberg H (1986) Effects of nitrogen deposition on the acidification of terrestrial and aquatic ecosystems. Water Air Soil Pollut 30:945–963

Grime JP (1979) Plant strategies and vegetation processes. John Wiley, Chichester, 222 pp

Guderian R, Küppers K, Six R (1985) Reaktionen von Fichte und Pappel auf Schwefeldioxid- und Ozon-Einwirkung bei unterschiedlicher Versorgung mit Kalzium und Magnesium. VDI Ber Düsseldorf 560:657–701

Gulmon SL, Mooney HA (1986) Costs of defense and their effect on plant productivity. In: Givnish T (ed) On the economy of plant form and function. Cambridge University Press, Cambridge, pp 681–695

Gundersen P, Rasmussen L (1988) Nitrification, acidification and aluminium release in forest soils. In: Nilsson J, Grennfelt P (eds) Critical loads for sulphur and nitrogen. Miljörapport 1988:15, Copenhagen, pp 225–268

Gundersen P, Rasmussen L (1990) Nitrification in forest soils. Effects from nitrogen deposition on soil acidification and aluminium release. Reviews of environmental contamination and toxicology 113:1–45. Springer, Berlin Heidelberg New York Tokyo

Gutschick VP (1981) Evolved strategies in nitrogen acquisition by plants. Am Nat 118:607–637

Hägglund B, Lundmark JE (1977) Site index estimation by means of site properties. Scots pine and Norway spruce in Sweden. Stud For Suec 138:1–38

Hain FP (1987) Interactions of insects, trees and air pollution. Tree Physiol 3:93–102

Harley JL, Smith SE (1983) Mycorrhizal symbiosis. Academic Press, London, 483 pp

Havill DC, Lee JA, Stewart GR (1974) Nitrate utilization by species from acidic and calcareous soils. New Phytol 73:1221–1231

Hawksworth DL, Rose F (1970) Qualitative scale for estimating sulphur dioxide air pollution in England and Wales using epiphytic lichens. Nature 227(5254):145–148

Heinselman ML (1970) Landscape evolution, peatland types and the environment in the Lake Agassiz Peatlands Natural Area, Minnesota. Ecol Monogr 40:235–261

Hesselman H (1917) Studien über die Nitratbildung in natürlichen Böden und ihre Bedeutung in pflanzenökologischer Hinsicht (Swedish with German summary). Medd Skogsförsöksanst Stockholm 13–14:297–528

Hesselman H (1937) Über die Abhängigkeit der Humusdecke von Alter und Zusammensetzung der Bestände im nordischen Fichtenwald von blaubeerreichen *Vaccinium*-Typ und über die Einwirkung der Humusdecke auf die Verjüngung und das Wachstum des Waldes (in Swedish with German summary). Medd Skogsförsöksanst Stockholm 30:529–716

Högberg P (1986) Soil nutrient availability, root symbioses and tree species composition in tropical Africa: a review. J Trop Ecol 2:359–372

Högberg P, Granström A, Johansson T, Lundmark-Thelin A, Näsholm T (1986) Plant nitrate reductase activity as an indicator of availability of nitrate in forest soils. Can J For Res 16:1165–1169

Holdgate MW, White GF (eds) (1977) Environmental issues. Scope Report 10. John Wiley, London, 224 pp

Holmbäck B, Malmström C (1947) Site-improvement experiments in lichen-pine forests in northern Sweden. Medd Stat Skogsforskn Inst Stockholm 36(6):89 pp (in Swedish with English summary)

Husar RB, Holloway JM (1983) Sulfur and nitrogen over North America. In: Ecological effects of acid deposition. SNV Rep 1636, Solna, Sweden, pp 95–115

Huss-Danell K, Lundmark JE (1988) Growth of nitrogen-fixing *Alnus incana* and *Lupinus* spp. for restoration of degenerated forest soil in northern Sweden. Stud For Suec 181:1–20

Hüttl RF (1985) „Neuartige" Waldschäden und Nährelementversorgung von Fichtenbeständen (*Picea abies* Karst.) in Südwestdeutschland. Freiburg Bodenkundl Abh 16:1–195

Hüttl RF, Zöttl HW (1985) Ernährungszustand von Tannenbeständen in Süddeutschland – ein historischer Vergleich. Allg Forst Z 1985 40(38):1011–1013

Ingestad T (1977) Nitrogen and plant growth: maximum efficiency of nitrogen fertilizers. Ambio 6:146–151

Ingestad T (1982) Relative addition rate and external concentration; driving variables used in plant nutrition research. Plant Cell Environ (1982):443–453

Ingestad T (1987) New concepts on soil fertility and plant nutrition as illustrated by research on forest trees and stands. Geoderma 40:237–252

Ingestad T, Arveby AS, Kähr M (1986) The influence of ectomycorrhiza on nitrogen nutrition and growth of *Pinus sylvestris* seedlings. Physiol Plant 68:575–582

Iversen J (1949) The influence of prehistoric man on vegetation. Danm Geol Unders 4. Raekke 3(6). CA Reitzel, Copenhagen, 25 pp

Jansen E, van Dobben HF (1987) Is the decline of *Cantharellus cibarius* in The Netherlands due to air pollution? Ambio 16:211–213

Jansson SL (1958) Tracer studies on nitrogen transformations in soil with special attention to mineralization-immobilization relationships. Ann R Agr Coll Sweden 24:101–361

Jansson SL (1987) Yield development in South Sweden with regard to natural conditions, cropping systems, and nutrient levels (Swedish with English summary). K Skogs- o Lantbruksakad Tidskr Suppl 19:9–20

Jenny H (1941) Factors of soil formation. McGraw-Hill, New York, 281 pp

Jenny H (1980) Soil genesis with ecological perspectives. Springer, Berlin Heidelberg New York, 560 pp

Johnsen I, Kielberg L, Kristiansen L, Mikkelsen TN, Riemer J, Ro-Poulsen H (1987) Effects of ozone, sulphur dioxide, nitrogen dioxide and simulated acid rain on nutrient content and water status of young *Picea abies*. In: Direct effects of dry and wet deposition on forest ecosystems – in particular canopy interactions. CEC Air Pollut Res Rep 4

Johnston AE, Mattingly GEG (1976) Experiments on the continuous growth of arable plots at Rothamsted and Woburn Experimental Stations: effects of treatments on crop yields and soil analyses and recent modifications in purpose and design. Adv Agron 27:927–956

Jordan CF (1985) Nutrient cycling in tropical forest ecosystems. John Wiley, Chichester, 190 pp

Keeney DR (1980) Prediction of soil nitrogen availability in forest ecosystems: a literature review. For Sci 25:159–171

Kenk G, Fischer H (1988) Evidence from nitrogen fertilisation in the forests of Germany. Environ Pollut 54:199–218

Kimmins JP, Scoullar KA (1984) The role of modelling in tree nutrition research and site nutrient management. In: Bowen GD, Nambiar EKS (eds) Nutrition of plantation forests. Academic Press, London, pp 463–487

Klemedtson L, Svensson BH (1988) Effects of acid deposition on denitrification and N_2O emission from forest soils. In: Nilsson J, Grennfelt P (eds) Critical loads for sulphur and nitrogen. Miljörapport 1988, Copenhagen, 15:343–362

Kononova M (1975) Humus of virgin and cultivated soils. In: Gieseking JE (ed) Soil components. I. Organic compounds. Springer, Berlin Heidelberg New York, pp 475–526

Krause GHM, Jung KD, Prinz B (1985) Experimentelle Untersuchungen zur Aufklärung der neuartigen Waldschäden in der Bundesrepublik Deutschland. VDI Ber Düsseldorf 560:627–656

Kreitinger JP, Klein TM, Novick NJ, Alexander M (1985) Nitrification and characteristics of nitrifying microorganisms in an acid forest soil. Soil Sci Soc Am J 49:1407–1410

Kriebitzsch WU (1978) Stickstoffnachlieferung in saure Waldböden Nordwestdeutschlands. Scr Geobot, Göttingen, 14:1–66

Lambert MJ (1986) Sulphur and nitrogen nutrition and their interactive effects on *Dothistroma* infection on *Pinus radiata*. Can J For Res 16:1055–1062

Lee JA (1986) Effects of NO_2 on aquatic ecosystems. In: Commission of the European Community Study on the need of a long-term limit value for NO_2. Brussels, pp 99–119

Lee JA, Stewart GR (1978) Ecological aspects of nitrogen assimilation. Adv Bot Res 6:1–43

Lee JA, Woodin SJ (1988) Vegetation studies and the interception of acid deposition by ombrotrophic mires. In: Verhoeven JTA, Heil GW, Werger MJA (eds) Vegetation structure in relation to carbon and nutrient economy. SPB Academic Publishing, The Hague, pp 137–147

Lemon E, van Houtte R (1980) Ammonium exchange at the land surface. Agron J 72:876–883

Lepp NW (1975) The potential of tree-ring analysis for monitoring heavy metal pollution patterns. Environ Pollut 9:49–61

Levi MP, Cowling EB (1969) Role of nitrogen in wood deterioration. VII. Physiological adaptation of wood-destroying and other fungi to substrates deficient in nitrogen. Phytopathology 59:460–468

Lind AM (1979) Nitrogen in soil water. Nord Hydrol 10:65–78

Linder S (1987) Responses to water and nutrients in coniferous ecosystems. In: Schulze ED, Zwölfer H (eds) Potentials and limitations in ecosystem analysis. Springer, Berlin Heidelberg New York Tokyo, pp 180–202

Linder S, Axelsson B (1982) Changes in carbon uptake and allocation patterns as a result of irrigation and fertilization in a young *Pinus sylvestris* stand. In: Waring RH (ed) Carbon uptake and allocation in subalpine ecosystems as a key to management. For Res Lab Oregon State Univ, pp 38–49

Linder S, Troeng E (1980) Photosynthesis and transpiration of 20-year-old Scots pine. In: Persson T (ed) Structure and function of northern coniferous forests – an ecosystem study. Ecol Bull Stockholm 32:165–181

Linkola M (1987) On the history of the rural landscape in Finland. Ethnologia Scandinavica 1987, Lund, pp 110–127

Lovelock JE (1979) Gaia, A new look at life on earth. Oxford University Press, Oxford

Lundmark JE (1977) The soil as part of the forest ecosystem (in Swedish). Sv Skogsvårdsför Tidskr 75:109–130

Madgwick HA, Tamm CO (1987) Allocation of dry weight increment in crowns of *Picea abies* as affected by stand nutrition. Oikos 48:99–105

Malmer N (1974) On the effects on water, soil and vegetation of an increasing atmospheric supply of sulphur. SNV PM 402E, 98 pp

Malmer N, Holm E (1984) Variations in the C/N quotient of peat in relation to decomposition rate and age determination with ^{210}Pb. Oikos 43:171–182

Malmström C (1949) Studien über Waldtypen und Baumartenverteilung im Län Västerbotten (in Swedish with German Summary). Medd Statens Skogsforskningsinst Stockholm 37(11):1–231

Mathy P (ed) (1988) Air pollution and ecosystems. Proceedings of an international symposium, Grenoble, France, 18–22 May 1987. Reidel, Dordrecht

Mattingly GEG, Chater M, Johnston AE (1975) Experiments made on Stackyard Field, Woburn, 1876–1974. III. Effects of NPK fertilizers and farmyard manure on soil carbon, nitrogen and organic phosphorus. Rep Rothamsted Expt Stn for 1974, Pt 2:61–77

Mattson S, Koutler-Andersson E (1954) Geochemistry of a raised bog. Lantbrukshögsk Ann 21:321–366

Mattson WJ (1980) Herbivory in relation to plant nitrogen content. Annu Rev Ecol Syst 11:119–161

Mattson WJ, Scriber JM (1987) Nutritional ecology of insect folivores of woody plants: nitrogen, fiber, and mineral considerations. In: Slansky F Jr, Rodriguez JG (eds) Nutritional ecology of insects, mites and spiders. John Wiley, New York

Matzner E (1988) Der Stoffumsatz zweier Waldökosysteme im Solling. Ber Forschungszentr Waldökosyst Univ Göttingen A 40, 217 pp

Mead DJ, Tamm CO (1988) Growth and stem form changes in *Picea abies* as affected by stand nutrition. Scand J For Res 3:505–513

Meijer K (1986) Critical loads for sulphur and nitrogen deposition in the Netherlands. In: Nilsson J (ed) Critical loads for nitrogen and sulphur. Nordisk ministerråd, Miljörapport 1986:11, Stockholm, pp 223–232

Melillo JM, Gosz JR (1983) Interactions of biogeochemical cycles in forest ecosystems. In: Bolin B, Cook RB (eds) The major biogeochemical cycles and their interactions. John Wiley, Chichester, pp 177–222

Melillo JM, Aber JD, Muratore JM (1982) Nitrogen and lignin control of hardwood leaf litter decomposition dynamics. Ecology 63:621–626

Melin E (1925) Untersuchungen über die Bedeutung der Baummykorrhiza. Eine ökologisch-physiologische Studie. G Fischer, Jena, 152 pp

Melin J, Nömmik H (1988) Fertilizer nitrogen distribution in a *Pinus sylvestris/Picea abies* ecosystem, central Sweden. Scand J For Res 3:3–15

Miller HG (1981) Forest fertilization: Some guiding concepts. Forestry 54:157–167

Miller HG (1984) Deposition-plant-soil interactions. Philos Trans R Soc Lond B 305:339–352

Miller HG, Miller JD (1988) Response to heavy nitrogen applications in fertilizer experiments in British forests. Environ Polut 54:219–232

Miller HG, Cooper JM, Miller JD, Pauline OJL (1979) Nutrient cycles in pine and their adaptation to poor soils. Can J For Res 9:19–26

Mitchell HL, Chandler RF (1939) The nitrogen nutrition and growth of certain deciduous trees of northeastern United States. With a discussion of the principles and practice of leaf analysis as applied to forest trees. Black Rock For Bull 11:1–94

Möller G (1983) Variation of boron concentration in pine needles from trees growing on mineral soil in Sweden, and response to nitrogen fertilizers. Commun Inst For Fenn 116:111–115

Müller PE (1887) Studien über die natürlichen Humusformen und deren Einwirkung auf Vegetation und Boden. (Translation of two papers in Danish, 1879 and 1884, where, as said in the preface to the 1887 paper, the terms "mull" and "mor" were scientifically defined; the German version uses "Torf" instead of "mor".) Springer, Berlin, 324 pp

Neftel A, Beer J, Oeschner H, Zürcher F, Finkel RC (1985) Sulphate and nitrate concentrations in snow from South Greenland. Nature 314:611–613

Nihlgård B (1985) The ammonium hypothesis – an additional explanation for the forest dieback in Europe. Ambio 14:2–8

Nilsson J (1986) (ed) Critical loads for nitrogen and sulphur. Miljörapport 1986: II. Copenhagen, 223 pp

Nilsson J, Grennfelt P (eds) (1988) Critical loads for sulphur and nitrogen. Miljörapport 1988: 15. Copenhagen, 418 pp

Nilsson SI, Berdén M, Popovic B (1988) Experimental work related to N deposition, nitrification and soil acidification – a case study. Environ Pollut 54:233–248

Nömmik H (1956) Investigations on denitrification in soil. Acta Agric Scand 6:195–228

Nömmik H, Möller G (1981) Nitrogen recovery in soil and needle biomass after fertilization of a Scots pine stand, and growth responses obtained. Stud For Suec 159:1–37

Nykvist N, Rosén K (1985) Effect of clear-felling and slash removal on the acidity of northern coniferous soils. For Ecol Manage 11:157–169

Nykvist N, Skyllberg U (1989) The spatial variation of pH in the mor layer of some coniferous forest stands in northern Sweden. Scand J For Res 4:3–11

Nylund B (1988) The regulation of mycorrhiza formation – carbohydrate and hormone theories reviewed. Scand J For Res 3:465–479

Odum HT (1957) Trophic structure and productivity of Silver Springs, Florida. Ecol Monogr 27:55–112

Olsen C (1921) Studies on the hydrogen concentration of the soil and its significance to the vegetation. CR Lab Carlsberg, Copenhagen, 15:1

Olson JS (1958) Rates of succession and soil changes on southern Lake Michigan sand dunes. Bot Gaz 119:125–170

Olson JS (1981) Carbon balance in relation to fire regimes. In: Mooney HA, Bonnichsen TM, Christensen NL, Lothan JA, Reiners WA (eds) Fire regimes and ecosystem properties US Dept Agr Techn Rep WO-26, Washington DC, pp 327–378

Parker G (1983) Throughflow and stemflow in the forest nutrient cycle. Adv Ecol Res 13:57–133

Pate JS (1980) Transport and partitioning of nitrogenous solutes. Annu Rev Plant Physiol 31:313–340

Pate JS (1983) Patterns of nitrogen metabolism in higher plants and their ecological significance. In: Lee JA, McNeill S, Rorison IH (eds) Nitrogen as an ecological factor. 22nd Symp Br Ecol Soc, Oxford, pp 225–256

Pate JS (1986) Economy of symbiotic nitrogen fixation. In: Givnish T (ed) On the economy of plant form and function. Cambridge University Press, Cambridge, pp 299–325

Persson H (1981) The effect of fertilization and irrigation on the vegetation dynamics of a pine-heath ecosystem. Vegetatio 46:181–192

Persson T, Bååth E, Clarholm M, Lundkvist H, Söderström BE, Sohlenius B (1980) Trophic structure, biomass dynamics and carbon metabolism of soil organisms in a Scots pine forest. Ecol Bull Stockholm 32:419–459

Popovic B (1966) Untersuchungen über den Stickstoffgehalt von Kiefernstammholz in einigen älteren Beständen in Schweden. Oikos 17:84–95

Popovic B (1977) Effect of ammonium nitrate and urea fertilizers on nitrogen mineralization, especially nitrification, in a forest soil. Research Notes, Departments of Forest Ecology and Forest Soils. R Coll For Stockholm 30:1–26

Raitio H, Rantala EM (1977) Macroscopic and microscopic symptoms of a growth disturbance in Scots pine. Description and interpretation (Finnish with summary in English). Commun Inst For Fenn 81:1–29

Raven JA (1983) Phytophages of xylem and phloem: a comparison of animal and plant sap-feeders. Adv Ecol Res 13:135–235

Read DJ (1983) The biology of mycorrhiza in the Ericales. Can J Bot 61:985–1004

Rehfuess KE (1981) Waldböden. Entwicklung, Eigenschaften und Nutzung. Parey, Hamburg, 193 pp

Rennie PJ (1955) Uptake of nutrients by mature forest trees. Plant Soil 7:49

Roberts TM, Darral NM, Lane P (1983) Effects of gaseous air pollutant on agriculture and forestry in the U.K. Adv Appl Biol 9:1–141

Roelofs JGM (1986) The effect of airborne sulphur and nitrogen deposition on aquatic and terrestrial heathland vegetation. Experientia 42:372–377

Roelofs JGM, Boxman AW, van Dijk HFG (1987) Effects of airborne ammonium on natural vegetation and forests. In: Asman WAH, Diederen HSMA (eds) Ammonia and acidification. Eurosap Symposium Proceedings, Bilthoven (The Netherlands), 266–276

Romell LG (1935) Ecological problems of the humus layer in the forest. Cornell Univ Agr Expt Sta Mem, Ithaca NY, 170:1–28

Romell LG (1967) Die Reutbetriebe und ihr Geheimnis. Stud Gen 6:362–369

Romell LG, Malmström C (1945) The experiments by H. Hesselman in lichen pine forest 1922–1942 (Swedish with English summary). Medd Statens Skogsförsöksanst Stockh 34:543–615

Rorison IH (1980) The effects of soil acidity on nutrient availability and plant response. In: Hutchinson TC, Havas M (eds) Effects of acid precipitation on terrestrial ecosystems. NATO Conference Series 1: Ecology 6:283–304. Plenum, New York

Rosén E (1982) Vegetation development and sheep grazing in limestone grasslands of south Öland, Sweden. Acta Phytogeogr Suec 72:1–108

Rosén K (1988) Effects of biomass accumulation and forestry on nitrogen in forest ecosystems. In: Nilsson J, Grennfelt P (eds) Critical loads for sulphur and nitrogen. Miljörapport 1988:15, Copenhagen, pp 269–294

Rosswall T (1983) Global balances of nitrogen and phosphorus. K Skogs-o Lantbruksakad Tidskr 122:287–292 (Swedish with English summary)

Rüger A, Prentice C, Owen M (1986) Results of the international waterfowl research bureau international waterfowl census 1967–1983. Population estimates and trends in selected species of ducks, swans and coot from the January counts in the Western Palaearctic. JWRB Spec Publ 6, Slimbridge

Rühling Å, Tyler G (1973) Heavy metal pollution and decomposition of spruce needle litter. Oikos 24:402–416

Rutter AJ (1957) Studies in the growth of young plants of *Pinus sylvestris* L. I. The annual cycle of assimilation and growth. Ann Bot Lond NS 21:399–426

Rydin H, Clymo RS (1989) Transport of carbon and phosphorus compounds about *Sphagnum*. Proc R Soc Lond B 237:63–84

Sakai A, Larcher W (1987) Frost survival of plants. Springer, Berlin Heidelberg New York Tokyo, 321 pp

Seip HM (1980) Acidification of freshwater – sources and mechanisms. In: Drablös D, Tollan A (eds) Ecological impact of acid precipitation. SNSF Project, Oslo-Ås, pp 358–366

Shaver GR, Chapin FS III (1980) Response to fertilization by various plant growth forms in an Alaskan tundra: nutrient accumulation and growth. Ecology 61:662–675

Shindler DW (1986) The significance of in-lake production of alkalinity. Water, air and soil Pollution 30:931–944

Sirén G (1955) The development of spruce forest on raw humus sites in northern Finland and its ecology. Acta For Fenn 62:4, 408 pp

Sjörs H (1954) Meadows in Grangärde Finnmark, SW Dalarna, Sweden. Acta Phytogeogr Suec 34:135 pp (Swedish with English summary)

Skeffington RA, Wilson EJ (1988) Excess nitrogen deposition: issues for consideration. Environ Pollut 54:159–184

Skinner FA, Boddey RM, Fendrik J (eds) (1989) Nitrogen fixation with non-legumes. Kluwer Acad Publ Dordrecht, 336 pp

Skye E (1968) Lichens and air pollution. A study of cryptogamic epiphytes and environment in the Stockholm region. Acta Phytogeogr Suec 52:1–123

Small E (1972) Photosynthetic rates in relation to nitrogen recycling as an adaptation to nutrient deficiency in peat bog plants. Can J Bot 50:2227–2233

Smirnoff N, Stewart GR (1985) Nitrate assimilation and translocation by higher plants: comparative physiology and ecological consequences. Physiol Plant 64:133–140

Söderlund R, Svensson BH (1976) The global nitrogen cycle. SCOPE Rep 7. Ecol Bull Stockh 22:23–73

Sollins P, Cromack K, Fogel R, Li CY (1981) Role of low-molecular-weight organic acids in the inorganic nutrition of fungi and higher plants. In: Wicklow DT, Carroll GC (eds) The fungal community. Marcel Dekker, New York, pp 607–619

Steen E (1980) Dynamics and production of semi-natural grassland vegetation in Fennoscandia in relation to grazing management. Acta Phytogeogr Suec 68:153–156

Stegemoeller K, Chappell HN (1989) Growth response to single and multiple applications of N fertilizer over 16 years in unthinned and thinned Douglas fir stands in the Pacific North West. Regional forest nutrition research project biennial rep 1986–1988. Coll For Resources Univ Wash Seattle Wash

Stevenson FJ (1982) Humus chemistry, genesis, composition, reactions. John Wiley, New York, 443 pp

Tamm CO (1953) Growth, yield and nutrition in carpets of a forest moss (*Hylocomium splendens*). Medd Stat Skogsforskn Inst Stockholm 43(1):140 pp

Tamm CO (1955) Studies on forest nutrition I. Seasonal variation in the nutrient content of conifer needles. Medd Stat Skogsforskn Inst Stockholm 45(5):34 pp

Tamm CO (1956) The response of *Chamaenerion angustifolium* (L.) Scop. to different nitrogen sources in water culture. Physiol Plant 9:331–337

Tamm CO (1964) Determination of nutrient requirements of forest stands. Int Rev For Res 1:115–170

Tamm CO (1974) Experiments to analyse the behaviour of young spruce forest at different nutrient levels. Proceedings of the First International Congress of Ecology, The Hague, Netherlands, September 8–14, pp 266–272

Tamm CO (1979) Nutrient cycling and productivity of forest ecosystems. In: Leaf AL (ed) Impact of intensive harvesting on forest nutrient cycling. Proceedings. State Univ New York Coll Environ Sci and Forestry. Syracuse, New York, pp 2–21

Tamm CO (1985) The Swedish optimum nutrition experiments in forest stands – aims, methods, yield results. K Skogs-o Lantbruksakad Tidskr Suppl 17:9–29 (Swedish with English summary)

Tamm CO (1989) Comparative and experimental approaches to the study of acid deposition effects on soils as substrate for forest growth. Ambio 18:184–191

Tamm CO, Hallbäcken L (1988) Changes in soil acidity in two forest areas with different acid deposition: 1920s to 1980s. Ambio 17:56–61

Tamm CO, Holmen H (1967) Some remarks on soil organic matter turn-over in Swedish podzol profiles. Medd Norske Skogforsöksvesen 85:69–88

Tamm CO, Popovic B (1974) Intensive fertilization with nitrogen as a stressing factor in a spruce ecosystem. I. Soil effects. Stud For Suec 121:1–32

Tamm CO, Popovic B (1989) Acidification experiments in young pine stands. SNV Reports No 3589, Stockholm

Tamm CO, Holmen H, Popovic B, Wiklander G (1974) Leaching of plant nutrients from forest soils as a consequence of forestry operations. Ambio 3:211–221

Tamm O (1950) Northern coniferous forest soils. Scrivener Press, Oxford, 253 pp

Thurston JM, Williams ED, Johnston AE (1976) Modern developments in an experiment on permanent grassland started in 1856: Effects of fertilisers and lime on botanical composition and crop and soil analyses. Ann Agron 27:1043–1082

Tilman D (1986) Nitrogen-limited growth in plants from different successional stages. Ecology 67:355–363

Tilman D (1987) Secondary succession and the pattern of plant dominance along experimental nitrogen gradients. Ecol Monogr 57:189–214

Tilman D (1988) Plant strategies and the dynamics and structure of plant communities. Monographs in Population Biology 26:360 pp

Toumey JW (1928) Foundations of silviculture upon an ecological basis. John Wiley, New York, 438 pp

Troedsson T (1952) Influence of geologic environment on the composition of the nutrients in solution in groundwater. Bull R School For Stockholm 10:1–16 (Swedish with English summary)

Troedsson T, Lyford WH (1973) Biological disturbance and small-scale spatial variations in a forested soil near Garpenberg, Sweden. Stud For Suec 109:1–23

Turner J (1977) Effects of nitrogen availability on nitrogen cycling in a Douglas fir stand. For Sci 23:307–316

Tyler G (1978) Leaching rates of heavy metal ions in forest soils. Water Air Soil Pollut 9:137–148

Tyler G (1987) Probable effects of soil acidification and nitrogen deposition on the floristic composition of oak (*Quercus robur* L.) forest. Flora (Berlin) 179:165–170

Uggla E (1957) Temperatures during controlled burning. The effect on the vegetation and the humus cover (in Swedish with English summary). Norrl Skogsv förb Tidskr 1957:443–500

Ulrich B (1983) A concept of forest ecosystem stability and of acid deposition as a driving force for destabilization. In: Ulrich B, Pankrath J (eds) Effects of accumulation of air pollutants in forest ecosystems. Reidel, Dordrecht, pp 1–29

Ulrich B, Mayer R, Khanna PK (1959) Deposition von Luftverunreinigungen und ihre Auswirkungen in Waldökosystemen im Solling. Schriften Forstl Fak Univ Göttingen 58:1–291

Unsworth MH, Ormrod DP (eds) (1982) Effects of gaseous air pollution in agriculture and horticulture. Proceedings Basler school Agric Science 32. Butterworth, London, 532 pp

van den Driessche R (1974) Prediction of mineral nutrient status by foliar analysis. Bot Rev 40:347–394

van Goor CP (1952) Deep cultivation and the production capacity of dry sandy forest soils. Uitvoer versl Boesbouwproefstation TNO 2:51–91, Wageningen (Dutch with English summary)

Verstraete W (1981) Nitrification. In: Clark FE, Rosswall T (eds) Terrestrial nitrogen cycles. Ecol Bull Stockh 33:303–314

Viereck LA (1970) Forest succession and soil development adjacent to the Chena river in interior Alaska. Arctic and Alpine Res 3:101–114

Vitousek PM (1982) Nutrient cycling and nutrient use efficiency. Am Nat 119:553–572

Vitousek PM (1983) The effect of deforestation on air, soil, and water. In: Bolin B, Cooke RB (eds) The major biogeochemical cycles and their interactions. SCOPE 21. John Wiley, New York, pp 223–245

Vitousek PM (1984) Litterfall, nutrient cycling, and nutrient limitation in tropical forests. Ecology 65:285–298

Vitousek PM, Andariese SW (1986) Microbial transformations of labelled nitrogen in a clearcut pine plantation. Oecologia (Berlin) 68:601–605

Vitousek PM, Matson PA (1984) Mechanisms of nitrogen retention in forest ecosystems: a field experiment. Science 225:51–52

Vitousek PM, Sanford RL Jr (1986) Nutrient cycling in moist tropical forest. Annu Rev Ecol Syst 17:137–167

Vitousek PM, Walker LR (1987) Colonization, succession and resource availability: ecosystem-level interactions. In: Gray AJ, Crawley MJ, Edwards PJ (eds) Colonization, succession and stability. 26th Symp Bri Ecol Soc, Oxford, pp 207–223

Vitousek PM, Gosz JR, Grier CC, Melillo JM, Reiners WA (1982) A comparative analysis of nitrification and nitrate mobility in forest ecosystems. Ecol Monogr 52:155–177

Waring RH, Franklin JF (1979) The evergreen coniferous forests of the Pacific North-West. Science 204:1380–1386

Waring RH, Pitman GB (1983) Physiological stress in lodgepole pine as a precursor for mountain pine beetle attack. Z Angew Entomol 96:265–270

Waring RH, Schlesinger WH (1985) Forest ecosystems. Concepts and management. Academic Press, Orlando, 340 pp

Wästerlund I (1982) Do pine mykorrhiza fungi disappear after fertilisation? (in Swedish with English summary). Svensk Bot Tidskr 76:411–417

Watson GA (1975) Special problems in tropical humid areas: soil and plant nutrient studies in rubber cultivation. In: FAO-IUFRO International Symposium on Forest Fertilization, Paris 1973

Weetman GF, Fournier RM (1984) Ten-year growth and nutrition effects of a straw treatment and of repeated fertilization on jack pine. Can J For Res 14:416–423

White DP, Leaf AL (1957) Forest fertilization, a bibliography. World Forestry Bull 2. Syracuse, New York, 305 pp

White TCR (1984) The abundance of invertebrate herbivores in relation to the availability of nitrogen in stressed food plants. Oecologia (Berlin) 63:90–105

WHO (1987) The effects of nitrogen on vegetation. In: Air quality guidelines for Europe. WHO Regional Publications, European Series Nr 23. Copenhagen, pp 373–385

Wijler J, Delwiche CC (1954) Investigations on the denitrifying process in soil. Plant Soil 5:155–169

Wittich W (1952) Der heutige Stand unseres Wissens vom Humus und neue Wege zur Lösung des Rohhumusproblems im Walde. Schriftenr Forstl Fak, Univ Göttingen 30:1–106

Wittig R, Ballach HJ, Brandt CJ (1985) Increase of number of acid indicators in the herb layer of the millet grass-beech forest of the Westphalian Bight. Angew Bot 59:219–252

Zackrisson O (1977) Influence of forest fires on the north Swedish boreal forest. Oikos 29:22–32

Zech W, Popp I (1983) Magnesiummangel, einer der Gründe für das Fichten- und Tannensterben in NO-Bayern. Forstwiss Centralbl 102

Zinke PJ (1980) Influence of chronic air pollution on mineral cycling in forests. US For Serv Pac SW For Range Expt Stn Gen Tech Rep 43:88–99

Zöttl H (1960) Methodische Untersuchungen zur Bestimmung der Mineralstickstoffanlieferung des Waldbodens. Forstwiss Centralbl 79:72–90

Subject Index

List of organizations who kindly permitted reproduction of material under their copyright:

Cambridge University Press	Fig. 2.1
Canadian Journal of Forest Research	Table 4.2
Ecological Bulletins, Stockholm	Fig. 2.2; 2.6
Elsevier Science Publishers Ltd.	Fig. 3.1
Pitagora Editrice Bologna	Fig. 3.6
SCOPE, Paris	Table 1.1
University of Chicago Press	Fig. 5.4
Verlag F. Golze, Göttingen	Fig. 5.5; 5.6